FM 6-02.53

TACTICAL RADIO OPERATIONS

August 2009

HEADQUARTERS, DEPARTMENT OF THE ARMY

Field Manual
No. 6-02.53

Headquarters
Department of the Army
Washington, DC
5 August 2009

TACTICAL RADIO OPERATIONS

Contents

Figures

Tables

Preface

This field manual (FM) serves as a reference document for tactical radio systems. (It does not replace FMs governing combat net radios, unit tactical deployment, or technical manuals [TMs] on equipment use.) It also provides doctrinal procedures and guidance for using tactical radios on the modern battlefield.

This FM targets operators, supervisors, and planners, providing a common reference for tactical radios. It provides a basic guidance and gives the system planner the necessary steps for network planning, interoperability considerations, and equipment capabilities.

This publication applies to Active Army, the Army National Guard (ARNG)/Army National Guard of the United States (ARNGUS), and the United States Army Reserve (USAR) unless otherwise stated. The proponent of this publication is the United States Army Training and Doctrine Command (TRADOC). The preparing agency is the United States Army Signal Center, approved by Combined Arms Doctrine Directorate. Send comments and recommendations on Department of the Army (DA) Form 2028 (Recommended Changes to Publications and Forms) directly to: Commander, United States Army Signal Center and Fort Gordon, ATTN: ATZH-IDC-CB (Doctrine Branch), Fort Gordon, Georgia 30905-5075, or via e-mail to signal.doctrine@conus.army.mil or signal.doctrine@us.army.mil.

Unless this publication states otherwise, masculine nouns and pronouns do not refer exclusively to men.

Chapter 1

Applications for Tactical Radio Deployment

This chapter addresses the Army's move to modularity and applications for tactical radio deployment from conventional corps to joint operations. It also includes a section on the Army Special Operations Forces (SOF) and the Army force generation process.

MODULARITY

1-1. The Army's transformation roadmap describes how the Army will sustain and enhance the capabilities of current forces while building future force capabilities to meet the requirements of tomorrow's joint force. It also describes how the Army will restructure the current force, creating modular capabilities and flexible formations while obtaining the correct mix between Regular Army and Army Reserve force structure. This rebalancing effort enhances the Army's ability to provide the joint team relevant and ready expeditionary land-power capability.

THE MODULAR ARMY CORPS AND DIVISION

1-2. The most significant advantage of modularization is greater strategic, operational, and tactical flexibility. The numbered Army Service component commander (ASCC), corps and division, will serve as—

- Theater's operational, strategic, and tactical command and control (C2).
- A land force and joint support element.
- C2 for a brigade combat team (BCT) or sustainment brigade, which serves as the primary tactical and support elements in a theater.

1-3. The modular numbered Army is organized and equipped primarily as an ASCC for a geographic combatant commander (GCC), or combatant command, and serves as the senior Army headquarters for an area of responsibility (AOR). It is a regionally focused, but globally networked, headquarters that consolidated most functions that were performed by the traditional Army and corps levels into a single operational echelon. The numbered Army is responsible for—

- Administrative control of all Army serviced personnel and installations in the GCCs AOR.
- Integrating Army forces into the execution of an AOR security cooperation plans.
- Providing Army support to joint forces, interagency elements, and multinational forces as directed by the GCC.
- Providing support to Army, joint, and multinational forces deployed to diverse joint operations areas.

1-4. The numbered Army modular design provides enough capability to execute an AOR entry and initial phases of an operation, while providing a flexible platform for Army and joint augmentation as the AOR develops. It provides administrative control of all Army personnel, units, and facilities in an AOR. The numbered Army is also responsible for providing continuous Army support to joint, interagency, and multinational elements as directed by the GCC, regardless of whether it is also controlling land forces in a major operation.

TACTICAL RADIO DEPLOYMENT

1-5. Tactical radios are deployed in support of the warfighting functions outlined in FM 3-0; movement/maneuver, fires, intelligence, sustainment, C2, and protection. The following paragraphs are an introduction of the tactical radio deployment throughout the Army to include BCTs and joint operations.

THEATER/ARMY

1-6. The theater/Army level is supported by signal companies within BCTs or expeditionary signal battalions (ESBs) depending on their mission and what type of support is needed. Some examples of combat net radio (CNR) communications that can be provided are—

- Single-channel tactical satellite (SC TACSAT).
- High frequency (HF) radio.
- Enhanced Position Location Reporting System (EPLRS) and EPLRS network (net) control capabilities.
- Single-Channel Ground and Airborne Radio System (SINCGARS) nets.
- Joint Network Node (JNN).

Note. For more information on theater/Army communications refer to Field Manual Interim (FMI) 6-02.45.

Note. As of June 2007 the Joint Network Node-Network program was incorporated into the Warfighter Information Network-Tactical (WIN-T) program and designated as WIN-T Increment 1. When JNN is used in this document it refers to the equipment and not to the program.

CORPS

1-7. C2 support at corps level is primarily provided by the integrated theater signal battalion (ITSB) or expeditionary signal battalion (ESB). The ITSB or ESB installs, operates, and maintains voice and data networks within and between corps C2 facilities. JNNs are the primary means to connect all elements of the corps and CNR networks perform a secondary role in the corps area of operations (AO). (For more information on JNN refer to FMI 6-02.60.)

DIVISION

1-8. Communications and information support at division level is provided by the division assistant chief of staff, command, control, communications, and computer operations (G-6) and the division signal company. The voice and data systems used by the division's AO are JNN, mobile regional hub nodes, tactical hub nodes, regional hub nodes, the tactical Internet, CNR nets, and the Global Broadcast Service.

1-9. The division signal company deploys JNN and tactical Internet networks in support of the division. The CNR systems deployed by the division are primarily SINCGARS, SC TACSAT, and HF radios. These systems are mostly user-owned and operated systems with the higher command responsible for net control.

BRIGADE

1-10. Communications and information support at maneuver brigade level is provided by internal brigade CNR assets. The SINCGARS, SC TACSAT, and HF radio are the primary means of communications within a maneuver brigade. The internal brigade signal company assets support C2 at brigade tactical operations centers (TOCs). Sustainment units operating in the division area behind the brigade sustainment area use CNRs as a secondary means of communications, with JNNthe brigade subscriber node, or mobile subscriber equipment (MSE) as the primary means of communications (some units that have not been fielded with JNN still have MSE).

BRIGADE COMBAT TEAMS

1-11. Communications and information support at the BCT level is provided by the brigade signal company. The brigade signal company is unique in structure and capabilities. It consists of the command and network operations sections, brigade support battalion, TOC nodal and the signal support platoons. The platoons support the BCT by providing—

- JNN.
- SC TACSAT.
- Brigade subscriber node that provides secure and non-secure voice, video, and data.
- EPLRS and EPLRS net control capabilities.
- Wireless network extension and capabilities.
- SINCGARS nets.

JOINT AND MULTINATIONAL OPERATIONS

1-12. Early planning and coordination are vital for reliable communications within the joint/multinational areas. Initial planning must be done at the highest level possible to ensure all contingency missions are included. Representatives from the host nation, multinational forces, and subordinate units should be present during coordination meetings; ensuring the individual requirements of multinational and subordinate commands is considered in the total communications plan. (Refer to Joint Publication [JP] 6-0 for additional information on joint communications planning and FM 6-02.72 for additional information on joint CNR issues.)

GEOGRAPHIC COMBATANT COMMANDER/ARMY SERVICE COMPONENT COMMANDER COMMUNICATIONS TEAM

1-13. The GCC/ASCC Communications Team provides communications support in the form of secure frequency modulation radio, UHF TACSAT, record telecommunications message support, and communications security (COMSEC) equipment maintenance to GCCs and/or ASCCs.

1-14. The GCC/ASCC Communications Team consists of—

- **Signal Systems Technician.** His duties are—
  - Supervises and manages the tactical Internet and administers the local area network and radio systems in TOC.
  - Plans, administers, manages, maintains, operates, integrates, secures, and troubleshoots Army Battle Command System (ABCS), Automated Information Systems (AIS), tactical data distribution, and radio systems.
  - Leads the team and personnel, and manages the training of personnel on the installation, administration, management, maintenance, operation, integration, securing, and troubleshooting of tactical ABCS/AIS, intranets, radio systems, and video teleconferencing systems.
  - Performs system integration and administration, and implements Information Assurance programs to protect and defend information, computers, and networks from disruption, denial of service, degradation, or destruction.
  - Develops policy recommendations and advise commanders and staffs on planning, installing, administering, managing, maintaining, operating, integrating, and securing ABCS/AIS, intranets, radio systems, and video teleconferencing systems on Army, Joint, Combined, and Multinational networks.
- **Electronic Systems Maintenance Technician.** His duties are—
  - Establishes team safety and crime prevention/security programs that adhere to the policies, practices, and regulations associated with these programs.
  - Manages personnel, equipment, and facility assets for operation, repair, maintenance, and modification of radio, radar, computer, electronic data processing, controlled cryptographic items, television, fiber optic, radiological and related communications equipment and associated tools, test, and accessory equipment.
  - Establishes team standing operating procedures (SOP) to ensure a proper work environment is maintained and that personnel adhere to maintenance schedules, the Army

Maintenance Management Systems, Quality Assurance and Quality Control procedures, and Standard Army Maintenance System-Level 1 (SAMS-1).

- Ensures personnel are trained to use the tools, test equipment, and applicable publications for the completion of the mission and are trained in automation skills.

- Ensures that the team is deployable by supervising the Unit Level Logistic System. Develops, rehearses, and implements load plans and deployment scenarios; establishes field SOPs; and ensures standards of the Mission Essential Task List are met.

- Ensures that Logistics tracking systems such as the Unit Level Logistic System, SAMS-2, and the Standard Army Retail Supply Systems are used. Interprets technical data and schematics, researches and interprets supply data, and fabricates repair parts or procures through outside resources. Coordinates technical, administrative, and logistical interface between the maintenance activity and supported units.

- Advises commander and staff on electronic equipment development, procurement, capabilities, limitations, and employment.

- Establishes, monitors, and maintains comprehensive environmental protection program IAW national and local directives.

- **Information Systems Chief** who is the principal information systems noncommissioned officer (NCO) for the GCC/ASCC Communications Team. His duties are—

  - Supervises, plans, coordinates, and directs the employment, operation, management and unit level maintenance of multi-functional/multi-user information processing systems in mobile and fixed facilities.

  - Provides technical and tactical advice to command and staff concerning all aspects of information processing system operations, maintenance and logistical support.

  - Supervises installation, operation, strapping, restrapping, preventive maintenance checks and services (PMCS) and unit level maintenance on COMSEC devices.

  - Conducts briefings on the status, relationship and interface of information processing systems within assigned area of interest.

  - Supervises or prepares technical studies, evaluations, reports, correspondence and records pertaining to multi-functional/multi-user information processing systems.

  - Plans, organizes and conducts technical inspections. Supervises development of the Information Systems Plan (ISP), Information Management Plan (IMP), and the Information Management Master Plan (IMMP).

  - Reviews, consolidates and forwards final written input for the Continuity of Operations Plan (COOP). Develops, enforces policy and procedure for facility Operations Security and physical security in accordance with regulations and policies.

  - Prepares or supervises the preparation of technical studies, evaluations, reports, correspondence, software programs, program editing, debugging and associated functions. Maintains records pertaining to information system operations.

- **COMSEC custodians.** They are responsible for—

  - Receipt, custody, security, accountability, safeguarding, inventory, transfer, and destruction of COMSEC material.

  - Supervision and oversight of hand-receipt holders to ensure compliance with existing COMSEC material security, accounting, operational policies/procedures, and acquisition, control, and distribution of all classified COMSEC material and cryptographic key in support of organizational missions.

- **Senior Information Technology NCO** who plans, supervises, coordinates, and provides technical assistance for the installation, operation, systems analyst functions, unit level maintenance, and management of multi-functional/multi-user information processing systems in mobile and fixed facilities. The Senior Information Technology NCO also—

  - Participates in development of the COOP, ISP, IMP and IMMP. Conducts quality assurance of information systems operations. Performs duties of COMSEC custodian in accordance with appropriate regulations. Supervises the operation of the Information Systems Security Officer (ISSO). Establishes and operates the printing and duplication program.

■ Supervises and implements classified document control policies, procedures, standards and inspections. Provides guidance on printing and publication account procedures, processes and regulatory requirements.

■ Controls production operations in support of command or agency priorities. Develop and enforce policy and procedures for facility management.

■ Develops, directs, and supervises training programs to ensure Soldier proficiency and career development. Organizes work schedules and ensure compliance with directives and policies on operations security, signal security, COMSEC and physical security.

■ Prepares or supervises the preparation of technical studies, evaluations, reports, correspondence and records pertaining to information system operations. Directs high level programming projects. Briefs staff and operations personnel on matters pertaining to information systems.

● **Information Technology NCO** who supervises the deployment, installation, operation, and unit level maintenance of multi-functional/multi-user information processing systems. His duties are—

■ Determines requirement, assign duties, coordinates activities of personnel engaged in information system analysis and maintenance.

■ Develops and administers on-site training programs. Compile output reports in support of information systems operations. Performs system studies using established techniques to develop new or revised system applications and programs. Analyzes telecommunications information management needs, and request logistical support and coordinate systems integration.

■ Ensures that spare parts, supplies, and operating essentials are requisitioned and maintained. Performs maintenance management and administrative duties related to facility operations, maintenance, security and personnel.

■ Performs COMSEC management functions and ISSO/Systems Administrator duties for the certification authority workstation. Prepares emergency evacuation and destruction plans for COMSEC facilities. Requisitions, receives, stores, issues, destroys and accounts for COMSEC equipment and keying material including over the air key.

■ Supervises ISSO functions. Provides verbal and written guidance and directions for the installation, operation and maintenance of specified battlefield information services.

■ Provides technical assistance; to resolve problems for information services in support personnel, functional users and functional staff.

● **Senior GCC/ASCC Communications NCO** who is responsible for supervising communications Soldiers of a GCC communications team. His duties are—

■ Supervises, plans and executes the installation, operation and maintenance of signal support systems, to include local area networks, wide-area networks and routers; satellite radio communications and electronic support systems; and network integration using radio, wire and battlefield automated systems.

■ Develops and implements unit level signal maintenance programs. Directs unit signal training and provides technical advice and assistance to commanders.

■ Develops and executes information services policies and procedures for supported organizations.

■ Coordinates external signal support mission requirements.

■ Prepares and implements Signal operations orders and reports.

■ Plans and requests Signal logistics support for unit level operations and maintenance.

- **GCC/ASCC Communications NCO.** His duties are—
  - Supervises, installs, maintains and troubleshoots signal support systems and terminal devices, to include radio, wire and battlefield automated systems.
  - Provides technical assistance and unit level training for automation, communication and user owned and operated automated telecommunications computer systems, to include local area networks and routers; signal communications support electronic equipment; and satellite radio communications equipment.
  - Disseminate information services policy, and prepares maintenance and supply requests for unit level signal support.
  - Operates and performs PMCS on assigned vehicles and on assigned power generators.

ARMY SPECIAL OPERATIONS FORCES

1-15. Army SOF includes the Special Forces, Ranger units, Special Operations Aviation Regiment, Civil Affairs (CA), and Psychological Operations (PSYOP).

SPECIAL FORCES

1-16. The Army Special Forces is organized into five active and two Army National Guard groups. In a tactical environment, Special Forces communications are strictly CNR. Special Forces units use the following CNR communication assets—

- Ultra high frequency (UHF) dedicated satellite communications (SATCOM) and demand assigned multiple access (DAMA).
- HF single side band (SSB), automatic link establishment (ALE), low probability of interception/detection (LPI/D), amplitude modulation (AM) and frequency modulation (FM) line of sight (LOS) radios.

1-17. The Special Operations Task Force has the capability to provide—

- Single-channel (SC) circuits (UHF DAMA and non-DAMA).
- HF SSB, ALE, and LPI/D.
- Very high frequency (VHF) and frequency modulation (FM) SINCGARS nets.
- Electronic mail (e-mail).
- Interface with the tactical Internet, MSE, and the Tri-Service Tactical Communications Program (if being utilized).

RANGERS

1-18. Ranger unit communications must be rapidly deployable and able to support airborne, air assault and infantry-type operations at all levels. Communications requirements are task organized to meet each mission's profile.

1-19. SC UHF SATCOM is the backbone of Ranger unit communications for links among headquarters, battalions, companies, and detachments. Other communications capabilities include:

- International maritime satellite (INMARSAT).
- UHF/VHF/FM/AM radios.
- HF SSB ALE.
- LPI/D.
- Multi-channel SATCOM augmentation may also be required.

SPECIAL OPERATIONS AVIATION REGIMENT

1-20. Special Operations Aviation Regiment communications provide air-to-air and air-to-ground aircraft communications for C2 mission deconflictions and mission support to SOF units. Air communications capabilities include:

- Multiband SATCOM.
- SC UHF SATCOM.
- HF burst and data.
- AM and FM radios.

1-21. Ground communications capabilities include: UHF SC SATCOM, HF, VHF or FM radio.

CIVIL AFFAIRS

1-22. SC SATCOM is the primary means of communications within CA units. While CA units receive other communications support from supported units or from commercial systems, they do have organic UHF/VHF/FM/AM, HF ALE and INMARSAT assets as well.

PSYCHOLOGICAL OPERATIONS FORCES

1-23. PSYOP communications support ensures the availability of communications and product distribution assets to PSYOP forces. Current and emerging technologies (military and commercial, including the PSYOP product distribution system and the Global Broadcast Service) will support the intelligence reach concept by providing secure, digital communications paths for transferring PSYOP products between the continental United States (CONUS) and deployed PSYOP units.

1-24. The PSYOP communications architecture consists of INMARSAT, SC TACSAT, and secure phones. Organic communications capabilities include SC UHF SATCOM, INMARSAT, HF, and FM radios.

ARMY FORCE GENERATION PROCESS

1-25. The Army force generation process creates three operational readiness cycles (reset/train pool, ready pool and available pool) where individual units increase their readiness over time, culminating into full mission readiness and availability to deploy. In order for signal Soldiers to be fully prepared once the unit reaches the available cycle they must have prior training on signal equipment. The following paragraphs address the importance of signal/CNR training during each cycle.

RESET/TRAIN POOL CYCLE

1-26. During the reset/train pool time it is important that leaders at all levels ensure that Soldiers are trained on current signal/CNR equipment. Some systems are more complex than others and require more familiarization.

1-27. It is during this cycle that new equipment training is also conducted by equipment fielding teams. It is important that new equipment is introduced as soon as possible so Soldiers have enough time to train and become proficient.

1-28. The unit must provide sustainment training to ensure individual skills do not decay and collective proficiency is attained to support mission accomplishment. New communications equipment, applications, and software updates are being fielded with greater frequency. Signal military occupational specialties are becoming more consolidated, and the highly specialized and technical skills required to operate communications systems are highly perishable.

READY POOL CYCLE

1-29. During the ready pool cycle Soldiers will receive critical training on signal equipment during sustainment training and field exercises (for example, the National Training Center, Maneuver Combat Training Centers, Joint Readiness Training Center and Combat Training Center.)

AVAILABLE POOL CYCLE

1-30. It is during the available pool cycle that units will be conducting deployments. Signal leaders should ensure Soldiers continue sustainment training on signal equipment as missions permit.

Chapter 2

Tactical Radios

This chapter provides an introduction to tactical radio operations. It addresses the tactical radio network, HF radios, VHF radios, UHF radios, SC TACSAT radios, airborne radios and other tactical radios being used. It also addresses electromagnetic spectrum operations (EMSO).

TACTICAL RADIO NETWORKS

2-1. The primary role of the network is voice transmission for C2. It assumes a secondary role for data transmission where other data capabilities do not exist.

2-2. Tactical communications networks change constantly. Unless control of the network is exercised, communications delay and a poor grade of service will result. The best method of providing this control without hampering operation is through centralized planning. Execution of these plans should be decentralized.

2-3. The planning and system control process helps communications systems managers react appropriately to the mission of the force supported, the needs of the commander, and the current tactical situation. The type, size, and complexity of the system being operated will establish the method of control.

2-4. Communications control is a process in which the matching of resources with requirements takes place. This process occurs at all levels of the control and management structure. In each case, the availability of resources is considered.

2-5. Operating systems control is the detailed hourly management of a portion of a theater Army, corps, or division communications system. Planning and control is according to the system being used.

2-6. The tactical radio network is designed around VHF radios (SINCGARS), HF radios, SC TACSAT and more recently, commercial off-the-shelf (COTS) radios are being used. Each system has unique and different capabilities and transmission characteristics that commanders consider to determine how to employ each system depending on the units' mission and other factors. (Refer to Appendix A for information on FM radio communication nets.)

HIGH FREQUENCY RADIOS

2-7. HF radios with ALE capability are replacing older HF systems. ALE permits radio stations to make contact with one another automatically. The success of ALE is dependent on effective frequency propagation and HF antenna construction and use.

VERY HIGH FREQUENCY RADIOS

2-8. SINCGARS is a family of VHF FM CNRs. They provide interoperable communications between surface and airborne C2 assets. SINCGARS has the capability to transmit and receive secure voice and data and is consistent with the North Atlantic Treaty Organization (NATO) interoperability requirements.

2-9. SINCGARS is secured with electronic attack (EA) security features (such as frequency hopping [FH]) that enable the United States (US) Army, United States Navy (USN), United States Air Force (USAF), and United States Marine Corps (USMC) communications interoperability. This interoperability ensures successful communications for joint and single component combat operations. (Refer to FM 6-02.72 for additional information regarding multi-service SINCGARS communications procedures.)

2-10. SINCGARS provides communications for units throughout the military. Data and facsimile transmission capabilities are available to tactical commanders through simple connections with various data terminal equipment (DTE).

2-11. The AN/PRC-148 and the AN/PRC-152 are COTS VHF LOS radios (multiband/multimode) that are being utilized in greater numbers in the Army today. The AN/PRC-148 was originally designed for the USMC and the SOF but the rest of the Army started to use the radio once its capabilities for small unit tactical operations were known. One of the features of multiband/multimode radios that is appealing to units is they all have SINCGARS and tactical satellite (TACSAT) capabilities.

ULTRA HIGH FREQUENCY RADIOS

2-12. UHF radios and systems play an import role in the military today. Radios such as the EPLRS, near term digital radio (NTDR), Multifunctional Information Distribution System (MIDS) and the Joint Tactical Information Distribution System (JTIDS) are being used throughout the Army for ground-to-air, ship-to-shore and multinational communications. UHF radios have been vital in recent urban combat situations.

SINGLE-CHANNEL TACTICAL SATELLITE

2-13. SC TACSAT systems provide another means for C2 communications in a tactical environment. SC TACSAT supports wideband and narrowband voice and data communications up to 64 kilobits per second (kbps) throughout the entire Army and Army SOF.

2-14. As more organizations take advantage of the range extension capabilities of SC TACSAT communications, there is potential for an overload in SATCOM. In response, the Army developed advanced SATCOM systems, such as the AN/PSC-5 (Spitfire), AN/PSC-5I (Shadowfire), and the AN/PSC-5D (multiband/multimode radio); and procured the AN/PRC-117F SC packable radios.

AIRBORNE RADIOS

2-15. Due to the nature of airborne operations, most of the radio systems used have air and ground capabilities or have ground and air versions to ensure that all elements of the tactical force have voice and data communications.

OTHER TACTICAL RADIOS

2-16. There are several other tactical radios and systems that are being used by units for different purposes. Handheld radios, such as the land mobile radio (LMR) and the integrated communications security (ICOM) F43G, are COTS radios being used by many units as platoon/squad radios for internal communications. There are also several survivor locator radios that are used by Special Forces, airborne and other units for search and rescue missions.

ELECTROMAGNETIC SPECTRUM OPERATIONS

2-17. EMSO is a core competency of the Signal Corps and falls under the purview of the signal staff officer (S-6)/G-6. In the Army, EMSO is performed by trained spectrum managers located in the S-6/G-6 from brigade to Army level. EMSO consists of planning, operating, and coordinating joint use of the electromagnetic spectrum through operational, planning, and administrative procedures. The objective of EMSO is to enable electronic systems to perform their functions in the intended environment without causing or suffering unacceptable frequency interference.

2-18. EMSO consists of four core functions; spectrum management, frequency assignment, host nation coordination, and policy. Through these core functions the spectrum manager uses available tools and processes to provide the Soldier with the spectrum resources necessary to accomplish the mission during all phases of operations. (For more information on EMSO refer to FMI 6-02.70.)

2-19. For CNR, the spectrum manager produces and distributes the corps units and division command level signal operating instructions (SOI) information. The corps spectrum manager assigns hopsets to corps units, restrictions to frequencies for hopset development, determines corps common hopsets, and allocates frequencies to the divisions for use in their hopsets and nets.

2-20. The corps SOI information is transferred to the divisions and from the division to the brigades for inclusion in their SOI data bases. This information is used to build the loadsets for the applicable radios (loadsets are the frequency data and COMSEC keys necessary for the radio to operate in FH mode).

This page intentionally left blank.

Chapter 3

High Frequency Radios

HF radios use ground and sky wave propagation paths to achieve short, medium, and long-range communications distances. The HF radio provides the tactical commander alternate means of passing voice and data communications. This chapter addresses the HF communications concepts, ALE, HF radios with ALE such as the AN/PRC-150 I and the improved high frequency radio (IHFR).

HIGH FREQUENCY COMMUNICATIONS CONCEPTS

3-1. The challenge of making HF radio systems work can be illustrated by contrasting them with the commonly used LOS radio systems. A well-designed, poorly-maintained LOS system will operate year after year with insignificant outages. On the other hand, even if the HF system is initially well designed, the HF radio-telephone operator (RTO) must continually adjust the system to compensate for the ionosphere, and an ever-changing terrestrial environment (interference from the other stations, atmospheric interference, and manmade noise).

3-2. Although HF radios are harder to maintain than the commonly used LOS radio, they provide a combination of simplicity, economy, transportability, and versatility that is impossible to match. For successful communications, radio frequency (RF) performance depends on—

- The type of emission.
- The amount of transmitter power output.
- The characteristics of the transmitter antenna. (To select the best antenna the planner must understand wavelength, frequency, resonance, and polarization. Antenna characteristics are addressed later in detail, in Chapter 9.)
- The amount of propagation path loss.
- The characteristics of the receiver antenna.
- The amount of noise received.
- The sensitivity and selectivity of the receiver.
- An approved list of usable frequencies within a selected frequency range.

3-3. The HF radio has the following characteristics that make it ideal for tactical long distance, wide area communication—

- HF signals can be reflected off the ionosphere at high angles that will allow beyond line of sight (BLOS) communications at distances out to 400 miles (643.7 kilometers [km]) without gaps in communications coverage.
- HF signals can be reflected off the ionosphere at low angles to communicate over distances of many thousands of miles.
- HF signals do not require the use of either SATCOM or wireless network extension assets.
- HF systems can be engineered to operate independent of intervening terrain or manmade obstructions.

3-4. Conducting tactical communications under urban combat/complex terrain conditions can be hard even for an experienced RTO. G-6/S-6 officers and radio planners need to know several factors that will provide the key to success—

- How to pick an antenna.
- Mode of transmission.

- Frequency band.
- Antenna masking.

3-5.   Training and implementing the units' HF equipment can help get messages through. Communications planners at every level need to understand the concepts of propagation, path loss, antennas, antenna couplers and digital signal processing. (Refer to Chapter 9 and Appendix B for more information on antennas and radio communications in unusual areas.)

AUTOMATIC LINK ESTABLISHMENT

3-6.   ALE is when a specialized radio modem, known as an ALE adaptive controller, is assigned the task of automatically controlling an HF receiver and transmitter, to establish the highest quality communications link with one or multiple HF radio stations. ALE controllers can be external devices or an embedded option in modern HF radio equipment.

3-7.   ALE controllers function on the basic principles of link quality analysis (LQA) and sounding (SOUND). These tasks are accomplished using the following common elements—

- Each controller has a predetermined set of frequencies (properly propagated for conditions) programmed into memory channels.
- Channels are continuously scanned (typically at a rate of two channels per second).
- Each controller has a predetermined set of net call signs programmed into memory that include its own station net call sign, net call signs, group call signs, and individual call signs.
- ALE controllers transmit LQA, which SOUND the programmed frequencies for best link quality factors on a regular, automated, or operator-initiated basis.
- When in a listening mode, ALE units (receiver/transmitter [RT]) log station call signs and associated frequencies, and assign a ranking score relevant to the quality of the link on a per channel basis.
- When a station desires to place a call, the ALE controller element attempts to link to the outstation using the data collected during ALE and SOUND activities. If the sending ALE has not collected the outstation's data, the controller will seek the station, and attempt to link a logical circuit between two users on a net that enables the users to communicate using all programmed channels.

3-8.   When the receiving station hears its address, it stops scanning and stays on that frequency. A handshake (a sequence of events governed by hardware or software, requiring mutual agreement of the state of the operational mode prior to information exchange) is required between the two stations. The two stations automatically conduct a handshake to confirm that a link was established. Upon a successful link, the ALE controllers will cease the channel scanning process, and alert the RTOs that the system has established a connection and that stations should now exchange traffic. Table 3-1 outlines communications between two stations during the handshake and LQA.

Table 3-1. ALE system handshake

	Call Station	Message	Receive Station
Handshake Process	B3B	"T6Y this is B3B"	T6Y
	Receive Station	Message	Call Station
	B3B	"B3B this is T6Y"	T6Y
	Call Station	Message	Receive Station
	B3B	"T6Y this is B3B"	T6Y
		Systems Linked	

3-9.   Table 3-2 outlines the LQA matrix for B3B. The channel numbers represent programmed frequencies, and the numbers in the matrix are the most recent channel-quality scores. Thus, if an RTO wanted to make a call from "B3B" to "T6Y", the radio would attempt to call on Channel 18, which has the highest LQA score.

3-10. When making multi-station calls, the radio (B3B) selects the channel with the best average score. Thus, for a multi-station call to all addresses in the matrix, Channel 14 would be selected.

Table 3-2. Notional link quality analysis matrix for a radio (B3B)

Address (call sign)	Channels				
	01	02	04	14	18
R3R	60	33	12	81	23
B6P	10	--	48	86	21
T6Y	--	--	29	52	63
E9T	21	00	00	45	--

3-11. Upon completion of a link session, the ALE controllers will send a link TERMINATION command, and return to the scanning mode to await further traffic. Built-in safeguards ensure that ALE controllers will return to the SCAN mode in case of a loss-of-contact condition.

3-12. Modern ALE controllers are capable of sending short orderwire digital messages known as automatic message displays to members of the net. Messages can be sent to any (ANY) or all (ALL) members of the NET or GROUP. ALE controllers can contact individual stations by their call sign, ALL stations or ANY stations on the NET or GROUP. ALL calls and ANY calls make use of wildcard characters in substitution for individual call signs such as @?@ (ALL) and @@? (ANY). NULL address calls are used for systems maintenance, and are sent as @@@. (For more information on HF ALE refer to FM 6-02.74.)

FREQUENCY SELECTION

3-13. For ALE to function properly, frequency selection is important. Consult with the frequency manager early on in the process. When selecting frequencies to use in a net, take into consideration the time of operation and distance to be communicated, power level and the type of antenna being used.

3-14. HF propagation changes daily. Lower frequencies work better at night and higher frequencies work better during the day. Frequencies need to be selected based on the type of network and the distance between radios.

3-15. When using the above parameters, a good propagation program should also be used to determine which frequencies will propagate. (Appendix C lists some of the propagation software programs available for use.)

THIRD GENERATION ALE

3-16. The third generation (3G) HF system uses a family of scalable burst waveform signaling formats for transmission of all control and data traffic signaling. Scalable burst waveforms are defined for the various kinds of signaling required in the system, to meet their distinctive requirements as to payload, duration, time synchronization, and acquisition and demodulation performance in the presence of noise, fading, and multipath. All of the burst waveforms use the basic binary PSK serial tone modulation at 2400 symbols per second that is also used in the military standard (MIL-STD) 188-110A serial tone modem waveform. The low-level modulation and demodulation techniques required for the new system are similar to those of the 110A modems.

3-17. In contrast to the MIL-STD-188-110A waveform, the waveforms used in the 3G HF system are designed to balance the potentially conflicting objectives of maximizing the time diversity achieved through interleaving, and minimizing on-air time and link turn-around delay. The latter objective plays an important role in improving the performance of ALE and automatic request for wireless network extension systems, which by their nature requires a high level of agility.

3-18. 3G ALE is designed to quickly and efficiently establish one-to-one and one-to-many (both broadcast and multicast) links. It uses a specialized carrier sense multiple access (CSMA) scheme to share calling channels, and monitors traffic channels prior to using them to avoid interference and collisions. Calling and traffic channels may share frequencies, but the system is likely to achieve better performance when they are separate. Each calling channel is assumed to be associated with one or more traffic channels that are sufficiently near in frequency to have similar propagation characteristics. The concept of associated control and traffic frequencies can be reduced to the case in which the control and traffic frequencies are identical.

3-19. 3G HF receivers continuously scan an assigned list of calling channels, listening for second generation (2G) or 3G calls. However, 2G ALE is an asynchronous system in the sense that a calling station makes no assumption about when a destination station will be listening to any particular channel. The 3G HF system includes a similar asynchronous mode; however, synchronous operation is likely to provide superior performance under conditions of moderate to high network load.

AN/PRC-150 I ADVANCED HIGH FREQUENCY/VERY HIGH FREQUENCY TACTICAL RADIO

Note. ALE HF radio systems procured by units are becoming more prevalent, as IHFRs such as AN/PRC-104, AN/GRC-192 and 213 are no longer in production. The ALE HF radio addressed in this section was recognized at publication time as being used in the field but not necessarily representative of all the ALE HF systems.

3-20. The AN/PRC-150 I radio, refer to Figure 3-1, provides units with state of the art HF radio capabilities in support of fast moving, wide area operations. HF signals travel longer distances over the ground than the VHF (SINCGARS) or UHF (EPLRS) signals do because they are less affected by factors such as terrain or vegetation. The AN/PRC-150 I and AN/VRC-104(V) 1 and (V) 3 vehicular radio systems, provide units with BLOS communications without having to rely on satellite availability on a crowded communications battlefield. The systems' manpack and vehicular configurations ensure units have reliable communications while on the move, and allow for rapid transmission of data and imagery.

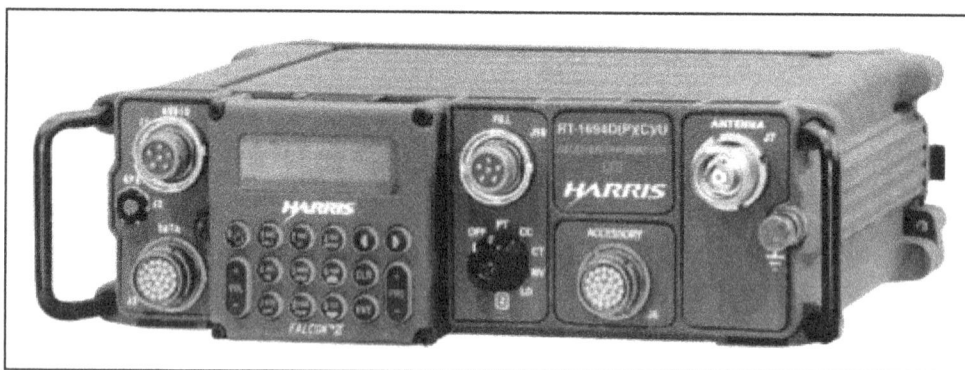

Figure 3-1. AN/PRC-150 I

3-21. The AN/PRC-150 I has the following characteristics and capabilities—

- Frequencies range from 1.6–29.9999 megahertz (MHz) using skywave modulation with selectable low, medium and high output power. It also operates from 20.0000–59.9999 MHz FM with a maximum output of 10.0 watts.
- Can be configured in manpack, mobile and fixed station configurations.
- Embedded Type I multinational COMSEC allows secure voice and data communications between ground and aircraft.
- Able to interface with SINCGARS cryptographic ignition key (CIK) is embedded in the removable key pad.
- Advanced electronic counter-countermeasures (ECCM) serial-tone FH improves communications reliability in jamming environments.
- Supports FH in HF narrowband, wideband and list.
- Programmable system presets for "one-button" operation.
- Internal tuning unit matches a wide variety of whip, dipole, and long-wire antenna automatically.
- Includes an internal, high-speed MIL-STD-188-110B serial-tone modem, which provides data operation up to 9,600 bits per second (bps).
- Embedded MIL-STD-188-141A ALE, digital voice 600 that simplifies HF operation by quickly and automatically selecting an accepted channel.
- Supports NATO Standardization Agreement (STANAG) 4538 automatic radio control system link set-up and data link protocols in 3G ALE radio mode.
- Supports networking capabilities using point-to-point protocol or Ethernet.
- Supports wireless Internet Protocol (IP) data transfer when operating in STANAG 4538 (3G).

3-22. The transceiver's extended frequency range (1.6–60 MHz) in combination with 16 kbps digital voice and data enables fixed frequency interoperability with other VHF FM CNRs. It provides Type 1 voice and data encryption compatible with advanced narrowband digital voice terminal (ANDVT)/KY-99, ANDVT/KY-100, VINSON/KY-57, and KG-84C cryptographic devices.

3-23. The AN/PRC-150 I is also capable of data communications by utilizing the TacChat software that is provided with the radio. Point-to-point data transmission can be completely secure and, with the use of the radios, 3G ALE synchronized scanning can be initiated quickly and smoothly.

MIXED EXCITATION LINEAR PREDICTION

3-24. Mixed excitation linear prediction (MELP) implemented in the AN/PRC-150 I can operate at both 600 and 2400 bps data rates. MELP has the ability to provide a significant increase in secure voice availability over degraded channels particularly at the 600 bps data rate when compared to other digital and analog forms of voice modulation.

3-25. The MELP speech mode uses an integrated noise pre-processor that reduces the effect of background noise and compensates for poor response at the lower speech frequencies. By using digital voice techniques such as band-pass filtering, pulse-dispersion filters, adaptive-spectral enhancement and adaptive noise pre-processing, voice communications performance over channels with low signal to noise (S/N) ratios typical of the urban combat environment can now be made useable and reliable.

3-26. The MELP capability is comparable to lowering the frequency, using higher power, and improving antenna efficiency which translates into decibels (dB) of "processing gain" and a better capability to communicate over urban terrain. In effect MELP is compensating for path loss and antenna inefficiency.

3-27. Last ditch voice (LDV) mode is designed to work when nothing else will. LDV takes advantage of digital voice processing at a much lower data rate (75 bps) in order to slash digital errors caused by marginal conditions. LDV is not a "real time" transmission mode but LDV has both a broadcast and an automatic-request for wireless network extension capability.

3-28. Voice data packets are created and sent in the transmitting radio. The radio then sends the packets at a very slow data rate using sophisticated error detection and correction digital coding techniques. Data packets are stored in the receiving radio and checked for errors in transmission caused by poor transmission path characteristics.

3-29. In an automatic request for wireless network extension mode corrupted packets can be returned to the transmitting radio in the event too many packets have too many errors for decoding into useable voice communications.

3-30. In broadcast mode all packets are stored upon receipt the first time. Radio software then assembles the packets and cues the RTO. The Soldier at the receiving radio then plays the message like a voicemail. The lower data rate and extensive signal processing can produce impressive performance since LDV can recover signals from below the noise levels. This can be equated to a considerable increase (3 dB or double) in transmitter power.

IMPROVING HIGH FREQUENCY RADIO OPERATIONS

3-31. According to the article "*Planning for the Use of High-Frequency Radios in the Brigade Combat Teams and Other Transforming Army Organizations*" whenever possible, man packed radios should be removed from the RTO's back and operated from the ground. This will decrease the capacitive coupling to ground effects of the RTO's body that reduce signal strength. The ground stake kit should also be connected to the radio terminal and driven into the earth when the radio is operated from the ground. The kit is provided with every radio and is designed to provide a low-resistance return path for ground currents. Using the kit dramatically improves signal strength and communications efficiency.

3-32. All antennas in the same net should also have the same polarization. Mixing polarization of antennas in a net as a rule will result in significant loss of signal strength due to cross polarization. The S-6 will therefore have to ensure that all stations in a net have the same (horizontal or vertical) antenna polarization when possible.

3-33. Signal strength can be improved by constructing radial wires to the ground. Radials need to be constructed from insulated wire and connected on one end to the radio ground terminal. The radials should be one-quarter wavelength long and secured to the earth on their ends by means of nails, stakes, etc. Distribution of the radials should be symmetrical. In operational terms for a brigade example, four wires (more if possible) of a practical length should be crossed in the center (X), and the center connected to the radio ground. The wires should be spread by 90 degrees and secured. (Chapter 9, Antennas, addresses how to construct a counterpoise which is similar to a radial.)

3-34. Using ground radials improves vertical antenna performance (gain) by allowing more current to flow in the antenna circuit and by lowering the antenna pattern's take off angle. This produces an increase in ground wave signal strength on low angles, where it is most useful for tactical communications. (Appendix D addresses radio operations in unusual environments.)

HF ANTENNA LOCATION CONSIDERATIONS

3-35. Units in a tactical fighting organization, when engaged in combat operations, will not always be able to locate their fixed and mobile radio assets at the most technically ideal positions for the best communications operations. HF communications planners should attempt to comply with as many of the following criteria as possible to gain the best technical advantages for the tactical situation—

- Use ground radials and ground stakes under vertical antenna to improve antenna efficiency and lower take off angles for better ground wave communications.
- Place vertical antennas on higher spots if possible, to enhance ground wave communications.
- Avoid placing vertical antennas behind metal fencing that will shield ground wave signals.
- Avoid placing vertical antenna near vertical conducting structures such as masts, tight poles, trees or metal buildings. Antennas need to be at distances of one wavelength or more to eliminate major pattern distortions and antenna impedance changes by induced current and reflections.

- Separate antenna as far as is practical to reduce interference effects between radio and antenna system. (For more information on HF radio operations in urban operations refer to "*AN/PRC-150 HF Radio in Urban Combat—a Better Way to Command and Control the Urban Fight.*")

IMPROVED HIGH FREQUENCY RADIOS

3-36. The IHFR is a SC, modular designed radio. It provides a versatile capability for short- and long-range communications. The capabilities of the IHFR make it flexible, securable, mobile, and reliable. However, the radio is the most detectable means of electronic communications, and is subject to intentional and unintentional electronic interference.

3-37. When using IHFRs, all transmissions will be secured with an approved cryptographic device (miniaturized terminal KY-99 or airborne terminal KY-100).

AN/PRC-104A MANPACK RADIO

3-38. The AN/PRC-104A, refer to Figure 3-2, consists of the RT-1209, amplifier/coupler AM-6874, antennas, and handsets. It is a low power radio which operates in the 2 to 29.999 MHz frequency range and passes secure C2 information over medium to long distances and varying degrees of terrain features that would prevent the use of VHF/FM CNR. It provides 280,000 tunable channels in 100 hertz (Hz) steps, and has automatic antenna tuning. (Refer to TM 11-5820-919-12 for more information on the AN/PRC-104A.)

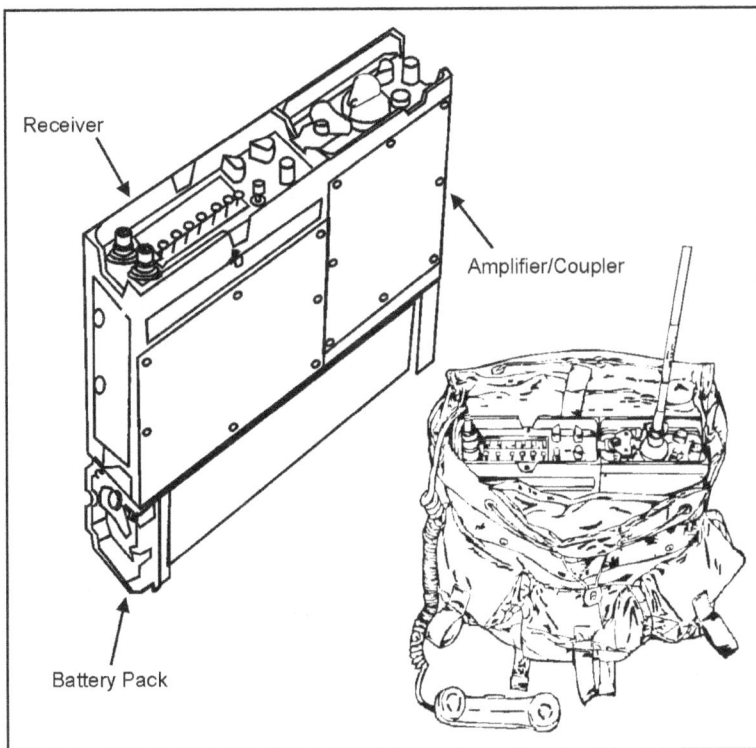

Figure 3-2. AN/PRC 104 manpack radio

LOW-POWER MANPACK/VEHICULAR RADIO, AN/GRC-213

3-39. The AN/GRC-213, refer to Figure 3-3, is a low power manpack/vehicular radio. It consists of the AN/PRC-104A radio, vehicle mount, amplifier power supply AM-7152, and three antennas (whip,

AN/GRA-50 doublet, and AS-2259 near-vertical incident sky wave [NVIS] antennas). Neither the AN/GRC-213 nor the AN/PRC-104A should be used for transmissions exceeding one minute within a 10 minute time frame. (Refer to TM 11-5820-923-12 for more information on the AN/GRC-213.)

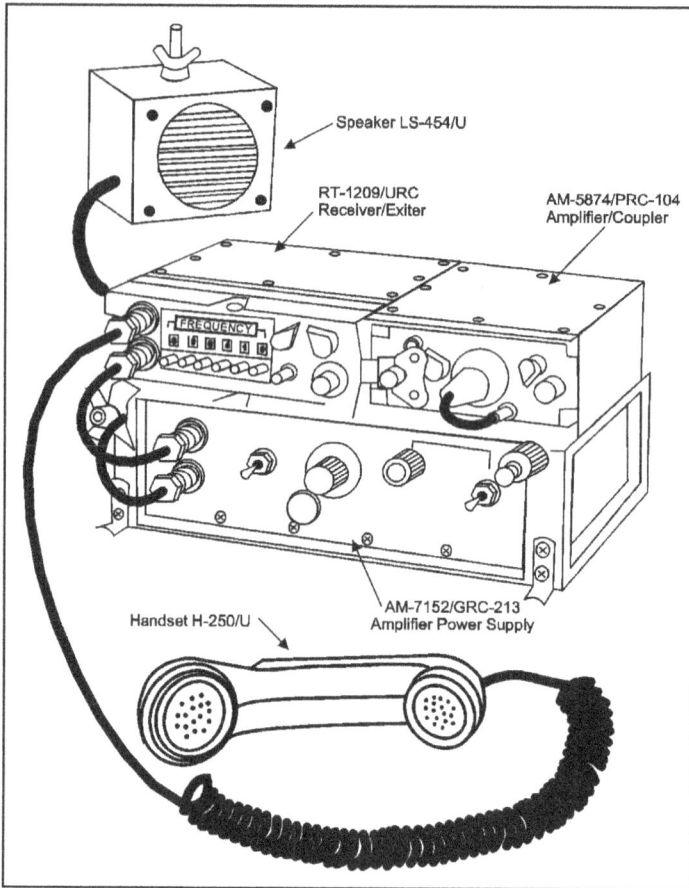

Figure 3-3. AN/GRC-213 low-power manpack/vehicular radio

HIGH-POWER VEHICLE RADIO, AN/GRC-193

3-40. The AN/GRC-193 (refer to Figure 3-4) is a medium/high power vehicular radio. The high power vehicular/airborne adaptive configuration consists of a basic RT (RT-1209) with required coupling device, amplifier, antenna (NVIS and whip antennas); data input/output (I/O) device; and external power sources. The radio will have the capability of selectable power (100 watts, 400 watts); normal operation will be at 100 watts. The AN/GRC-193 uses the KY-99 for securing voice traffic, and uses the telecommunications security (TSEC)/KG-84 for securing data traffic. The antenna may be remoted up to 61 meters (200 feet [ft]) from the radio set, using the antenna siting built-in test (BIT) that is part of the basic configuration. (Refer to TM 11-5820-924-13 for more information on the AN/GRC-193.)

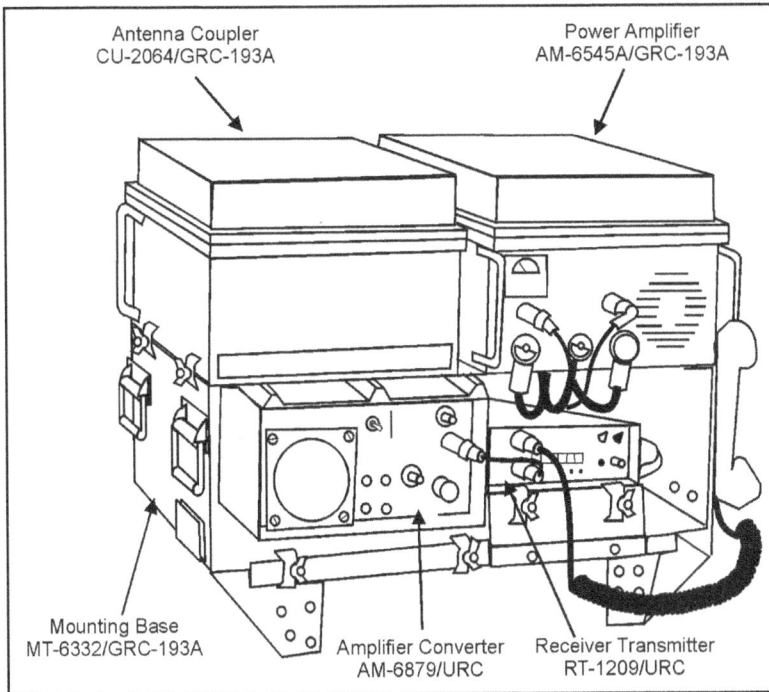

Figure 3-4. AN/GRC-193 high-power vehicle radio

This page intentionally left blank.

Chapter 4

Very High Frequency Radio Systems

SINCGARS provide interoperable communications between C2 assets and have the capability to transmit and receive secure voice and data. This chapter describes the SINCGARS, its components, enhancements, and ancillary equipment. It also addresses SINCGARS planning, secure devices, VHF FM wireless network extension stations and SINCGARS jamming and anti-jamming. Due to the high usage of COTS radios, other VHF radios included are the AN/PRC-148 and AN/PRC-152.

SINGLE-CHANNEL GROUND AND AIRBORNE RADIO SYSTEM CHARACTERISTICS AND CAPABILITES

4-1.   The SINCGARS family is designed on a modular basis to achieve maximum commonality among various ground and airborne configurations. A common RT is used in the manpack and all vehicle configurations. These individual components are totally interchangeable from one configuration to the next. Additionally, the modular design reduces the burden on the logistics system to provide repair parts.

4-2.   SINCGARS operates in either the SC or FH mode. It is compatible with all current US and multinational VHF radios in the SC non-secure mode. SINCGARS is compatible with other USAF, USMC, and USN SINCGARS in the FH mode. SINCGARS stores eight SC frequencies, including the cue and manual frequencies and six separate hopsets.

4-3.   SINCGARS operates on any of 2,320 channels between 30–88 MHz, with a channel separation of 25 kilohertz (kHz). It is designed to operate in nuclear or hostile environments.

4-4.   SINCGARS accepts either digital or analog input and imposes the signal onto a SC or FH output signal. In FH, the input changes frequency about 100 times per second over portions of the tactical VHF range. This hinders threat intercept and jamming units from locating or disrupting friendly communications.

4-5.   SINCGARS provides data rates of 600, 1,200, 2,400, 4,800, and 16,000 bps; enhanced data mode (EDM) of 1200N, 2400N, 4800N, and 9600N; and packet and recommended standard-232 data. The system improvement program (SIP) and advanced system improvement program (ASIP) radios provide EDM, which provide forward error correction (FEC), speed, range, and data transmission accuracy.

4-6.   SINCGARS has the ability to control output power. The RT has three power settings that vary transmission range from 200 meters (656.1 ft) to 10 km (6.2 miles). Adding a power amplifier (PA) increases the LOS range to 40 km (25 miles). The variable output power level allows users to lessen the electromagnetic signature given off by the radio set.

4-7.   Using lower power is particularly important at major command posts (CPs), which operate in multiple networks. The ultimate goal is to reduce the electronic signature at the CPs. The net control station (NCS) should ensure all members of the network operate on the minimum power necessary to maintain reliable communications.

4-8.   SINCGARS also has BIT functions that notify the RTO when the RT is malfunctioning. It also identifies the faulty circuits for repair or maintenance.

4-9.   SINCGARS provides outside network access through a hailing method. The cue frequency provides the hailing ability to the SINCGARS. When hailing a net, an individual outside the net contacts the alternate NCS on the cue frequency. The NCS must retain control of the net. Having the alternate NCS go to the cue assists in managing the net without disruption. In the active FH mode, the SINCGARS gives audible and visual signals to the RTO that an external subscriber wants to communicate with the FH net. The SINCGARS alternate NCS RTO must change to the cue frequency to communicate with the outside radio system.

4-10. The net uses the manual channel for initial network activation. The manual channel provides a common frequency for all members of the net to verify the equipment is operational. During initial net activation, all RTOs in the net tune to the manual channel using the same frequency. After establishing communications on the manual channel, the NCS transfers the hopset variables to the out stations and then switches the net to the FH mode.

4-11. The NCS is responsible for—

- Opening and closing a net.
- Maintaining net discipline.
- Controlling net access.
- Knowing who is a member of the net.
- Imposing net controls.

4-12. Refer to Appendix B for more information on SC radio communications principles, Chapter 12 for proper radio procedures, Appendix E for Julian date, sync time and time conversion and TM 11-5820-890-10-5 for more information on SINCGARS NCS.

SINGLE-CHANNEL GROUND AND AIRBORNE RADIO SYSTEM RADIO SETS

4-13. Using common components in SINCGARS is the key to tailoring radio sets for specific missions with the RT being the basic building block for all radio configurations. The number of RTs, amplifiers, the installation kit, and the backpack component determine the model. Table 4-1 compares the components of several versions of SINCGARS. For more information on SINCGARS refer to TM 11-5820-890-10-5, TM 11-5820-890-10-8 and Technical Bulletin (TB) 11-5821-333-10-2.

Table 4-1. Comparison of SINCGARS versions and components

	Short Range (consist of 1 radio)	Long Range (consist of 1 radio)	PA	Dismount Manpack	Vehicular Amplifier Adapter (VAA) (AM-7239C/E)
AN/VRC-87	X				X
AN/VRC-88	X			X	X
AN/VRC-89	X	X	X		X
AN/VRC-90		X	X		X
AN/VRC-91	X	X	X	X	X
AN/VRC-92		X (2)	X (2)		X
AN/PRC-119	X			X	

4-14. There are several ground unit versions of SINCGARS (RT-1523/A/B/C/D/E) and three airborne versions (RT-1476/1477/1478). Most airborne versions require external COMSEC devices. The RT-1478D has ICOM and an integrated data rate adapter (DRA). (Airborne SINCGARS versions are addressed in Chapter 7.)

4-15. Airborne and ground versions are interoperable in FH and SC operations. The airborne versions differ in installation packages and requirements for data capable terminals.

GROUND VERSION RECEIVER/TRANSMITTER

4-16. Either the RT-1523/A/B/C/D (refer to Figure 4-1) or the RT-1523E (refer to Figure 4-2) comprise the core component of all ground-based radio sets. The RT-1523 series has internal COMSEC circuits (source of the ICOM designation). The ground versions are equipped with a whisper mode for noise restriction during patrolling or while in defensive positions. The RTO whispers into the handset and is heard at the receiver in a normal voice.

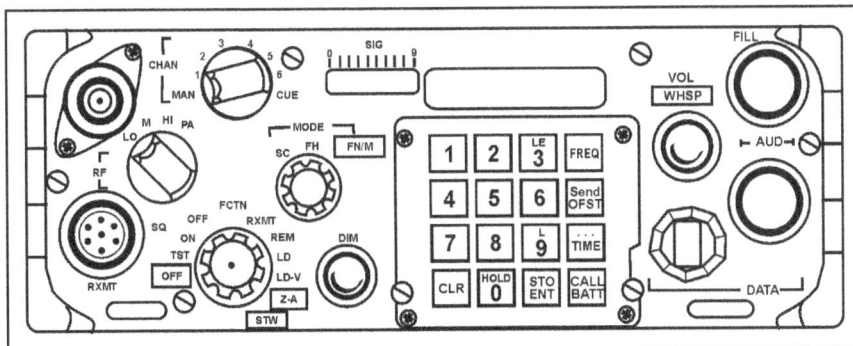

Figure 4-1. Front panel ICOM radio RT-1523/A/B/C/D

Figure 4-2. Front panel ICOM radio RT-1523E

ADVANCED SYSTEM IMPROVEMENT PROGRAM

4-17. The SINCGARS ASIP increases the performance of the SINCGARS SIP (RT-1523 C/D models). It also increases its operational capability in support of the tactical Internet, specifically improved data capability, manpower and personnel integration requirement compliance, and flexibility in terms of interfaces with other systems. Figure 4-3 is an example of the SINCGARS ASIP radio.

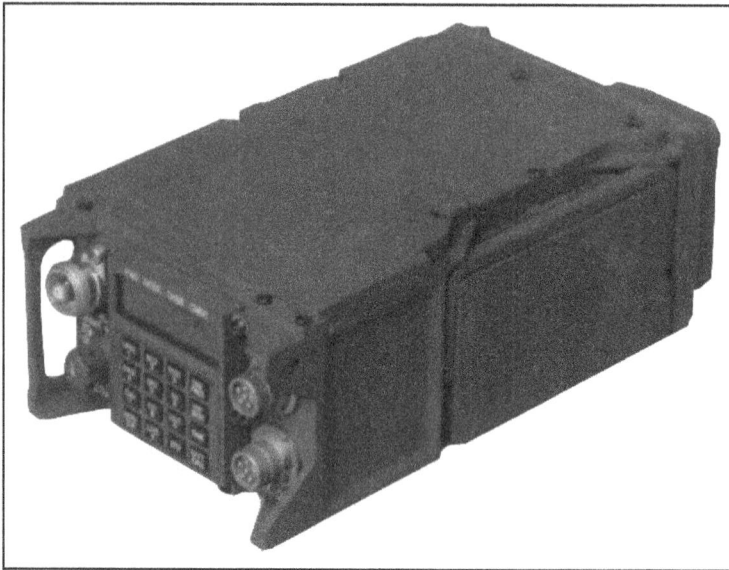

Figure 4-3. SINCGARS ASIP radio

4-18. Table 4-2 outlines a comparison of the SINCGARS ICOM, SINCGARS SIP, and the SINCGARS ASIP. All ASIP radios can be physically remoted by another ASIP radio up to 4 km (2.4 miles) away, via a two-wire twisted pair (typically WD-1 or WF-16). To remote a radio, an external two-wire adapter is used as the interface between the radio and the wires. This remote control feature can be performed between the dismounted RT and the VAA, or between two dismounted RTs. Another host controller can control the ASIP radio via the external control interface when the ASIP radio system is integrated as part of a larger system.

Table 4-2. SINCGARS enhancements comparison

ICOM capabilities (RT-1523A/B) Point-to-point communications	SIP capabilities (RT-1523C/D) Point-to-point communications	ASIP capabilities (RT 1523E/F) Point-to-point communications
1. FH per MIL-STD-188-241.	1. FH per MIL-STD-188-241.	1. Same as SIP.
2. SC per STANAG 4204.	2. SC per STANAG 4204.	
3. Mode 1, 2, 3 fill.	3. Mode 1, 2, 3 fill.	
4. Electronic remote fill (ERF).	4. ERF.	
Plain text (PT) and cipher text (CT) mode	Circuit switching and packet network communications	Circuit switching and packet network communications
1. Railman COMSEC.	1. CSMA protocol.	1. Same as SIP.
2. Seville advanced remote keying.	2. Railman COMSEC.	
	3. Seville advanced remote keying.	

Table 4-2. SINCGARS enhancements comparison (continued)

Point-to-point data communications	Point-to-point data communications	Point-to-point data communications
1. 600 to 4,800 bps standard data mode. 2. Tactical Fire Direction System (TACFIRE), analog data. 3. Transparent 16 kbps data.	1. 600 to 4,800 bps standard data mode. 2. TACFIRE, analog data. 3. Transparent 16 kbps data. 4. 1,200 to 9,600 bps EDM data. 5. Recommended standard-232 EDM data. 6. Packet data. 7. External control interface.	1. Same as SIP.
Other features	*Other features*	*Other features*
1. Noisy channel avoidance. 2. Enhanced message completion.	1. Noisy channel avoidance. 2. Enhanced message completion. 3. External global positioning system (GPS) interface. 4. Embedded GPS hooks. 5. Remote control unit (RCU).	1. Same as SIP plus— • Enhanced system improvement program (ESIP) waveform. • Faster channel access to reduce net fragmentation. • Enhanced noisy channel avoidance algorithm to improve FH sync probability. • Improved time of day tracking and adjustments. • Extra end of message hops to improve sync detection and reduce fade bridging. • Embedded battery.
VAA (AM-7239B):	*VAA (AM-7239C):*	*VAA (AM-7239E):*
1. Dual transmit power supply.	1. Dual transmit power supply. 2. Host interface. 3. Backbone interface. 4. MIL-STD-188-220A.	1. Same as SIP plus— • More powerful 860 microprocessor. • Ethernet interface. • Enhanced protocols. • Increased memory and buffer size.

Enhanced System Improvement Program Capabilities

4-19. The SINCGARS ASIP radio incorporates an ESIP waveform. The waveform includes optimizations to the algorithms of the noisy channel avoidance scheme, the time of day tracking scheme, and the end of message scheme. Enhancements include—

- **ESIP waveform**—implements a faster channel access protocol, which reduces net fragmentation by shortening the collision intervals between voice and data transmissions. The result is the reduction of voice and data contention problems associated with shared voice and data networks.
- **Noisy channel avoidance algorithm**—always reverts to a known good frequency instead of constantly searching for clear frequencies, thus increasing the FH synchronization probability in high noise and jamming conditions.
- **Time of day enhancement**—uses a reference BIT that assures time constraints are the same during each transmission.
- **End of message enhancement**—reduces fade bridging, whereby the transmission would linger even though adding extra end of message hops to increase the detection and probability of synchronization completes the message.

SINCGARS INTERNET CONTROLLER CARD

4-20. The internet controller card (INC) was introduced as part of the SINCGARS VAA to support the seamless flow of data across the battlefield, permitting both horizontal and vertical flow of C2 information. The INC is in the right hand side of the VAA, and is only needed when the SINCGARS system is operating in the packet mode of operation. Figure 4-4 is an example of a VAA.

4-21. The packet mode allows for the sharing of voice and data over the same operational net. A store and forward feature in the INC delays data while voice traffic is ongoing, and puts data on the net when the push-to-talk is released for voice. When the INC is loaded with initialization data, it will contain routing tables that identify the addresses of all members with which it is affiliated, as well as other radio nets that it can route to. The host computers generate messages, along with the IP addresses of the individual(s) to whom it is being sent.

4-22. When a message reaches an INC, the INC looks up its routing table to determine whether that message is for a member of its net or whether it needs to be sent off to the next adjoining net. The packet mode will automatically continue this routing process until it reaches its destination. The packet mode knows if the message is for someone within its net, and if the message stops there it will not get wirelessly networked extended out. This differs in a wireless network extension site, in that everything received at the wireless network extension station is relayed.

4-23. The VAA mounted INC is the predominant communications router for the tactical maneuver platforms participating in a SINCGARS enabled tactical Internet. The INC routes data between SINCGARS and EPLRS. The INC uses commercial IP services to deliver unicast and multicast data packets that consist of C2 and situational awareness (SA) messages.

4-24. The INC has an improved microprocessor with increased memory buffer size and an Ethernet interface is also available. Access to the Ethernet interface is through the same 19 pin connector used for the EPLRS interface. Two of the nineteen pins are used as twisted pairs to provide for the 10Base-T Ethernet connection. This feature will allow multiple INCs to be connected for the sharing or dissemination of information in a local area network configuration (such as in a TOC environment).

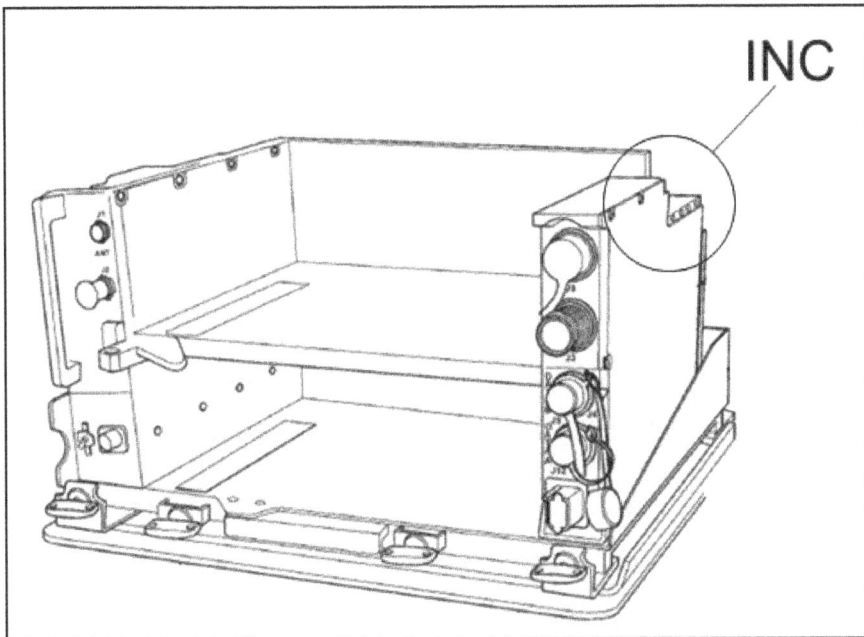

Figure 4-4. Vehicular amp adapter and INC

SINGLE-CHANNEL GROUND AND AIRBORNE RADIO SYSTEM ANCILLARY EQUIPMENT

4-25. Remote control devices, data fill/variable storage transfer devices, and the vehicular intercommunications system (VIS) are the main categories of ancillary equipment associated with SINCGARS addressed in the following paragraphs.

4-26. Remote control devices are divided into intra-vehicular and external remotes. The intravehicular remote control unit (IVRCU), C-11291, is the remote for intra-vehicular radio control. The securable remote control unit (SRCU), C-11561, is used to remote radios off the main site location. Additionally, the SIP/ASIP radio can be used as a RCU by merely selecting the RCU option under the RCU key of the SIP/ASIP RT keypad.

INTRAVEHICULAR REMOTE CONTROL UNIT

4-27. The IVRCU, C-11291 can be used with either an ICOM or non-ICOM radio. It can control up to two mounting adapters with up to three separate radio sets from a single station. The IVRCU can also be connected in parallel so that two different RTOs, such as the vehicle commander and the vehicle driver, can control the radios from their respective positions in the vehicle. The radio function switch must be set in the remote operating position for the external control monitor to function correctly. Refer to Figure 4-5 for more information on Intravehicular remote control unit, C-11291.

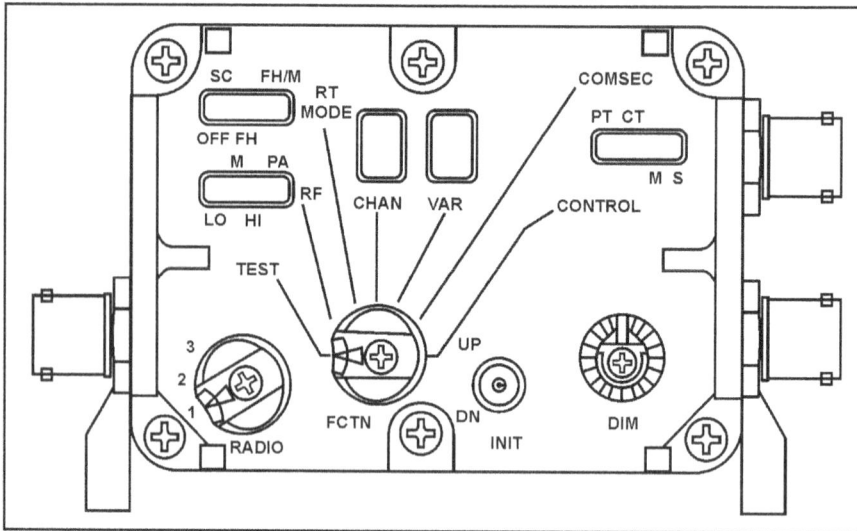

Figure 4-5. Intravehicular remote control unit, C-11291

SECURABLE REMOTE CONTROL UNIT

4-28. The SRCU, C-11561 can securely remote a single radio up to 4 km (2.4 miles). The SCRU and the RT are connected using field wire on the binding posts of the amplifier adapter or battery box. The SRCU appears and operates almost identically to the RT. The SRCU can secure the wire line between the radio and the terminal set. The SRCU controls all radio functions including power output, channel selection, and radio keying (refer to Figure 4-6, Securable remote control unit, C-11561).

4-29. The remote also provides an intercom function from the radio to the terminal unit and vice versa. The COMSEC and data adapter devices may be attached directly to the SRCU for secure communications over the transmission line, and optimal interface with digital data terminals. The SCRU replaced the AN/GRA-39. Four main configurations of the SRCU include—
- Manpack; radio in vehicular mounting adapter.
- Vehicular mounting adapter; radio in manpack.
- Manpack; radio in manpack.
- Vehicular mounting adapter; radio in vehicular mounting adapter.

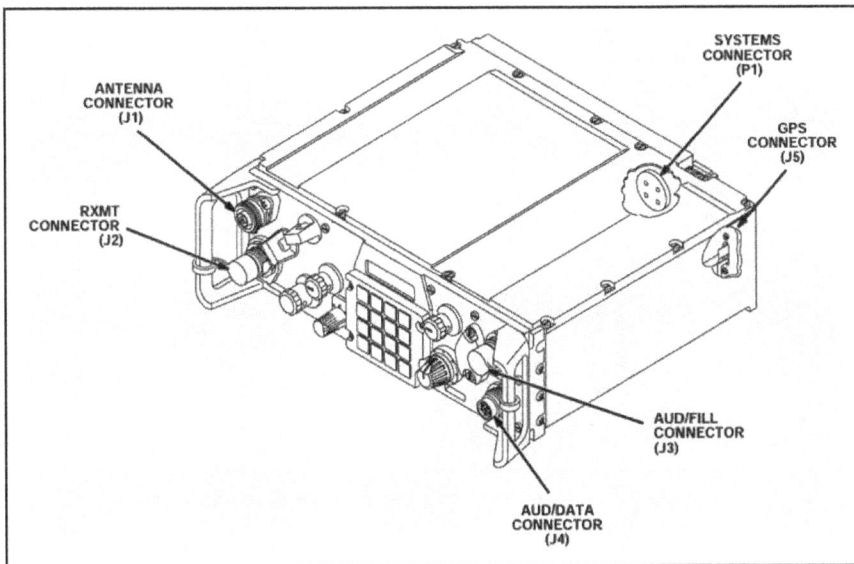

Figure 4-6. Securable remote control unit, C-11561

AN/CYZ-10, AUTOMATED NET CONTROL DEVICE

4-30. The AN/CYZ-10, automated net control device (ANCD) is capable of receiving cryptographic net information from the Army Key Management System (AKMS) workstation. It can obtain keys from the system key generators or from a hard copy key. Once received, the keys are correctly matched to the cryptographic net information.

4-31. The ANCD has the capacity to store a large number of keys along with related information that will assist a cryptographic NCS in accounting for, distributing, updating, and replacing cryptographic keys. Figure 4-7 is an example of the ANCD, AN/CYZ-10. (Refer to TB 11-5820-890-12 for more information on the ANCD.)

4-32. The ANCD is primarily used for handling COMSEC keys, FH data, sync times, and SOI information. A typical ANCD data load at the operator level consists of two loadsets (COMSEC keys and FH data for all six radio channels), each is good for 30 days of operation, plus 60 days of SOI information, structured in five ten-day editions, containing two five-day sets each. The ANCD eliminates the need for most paper SOI products. The ANCD replaces the KYK-13, KYX-15, MX-18290, and the MX-10579 in support of SINCGARS.

4-33. The ANCD can store up to 20 loadsets (COMSEC and FH data). The number of smaller unit SOI editions that can be stored in an ANCD depends on the size of the SOI extract. The ANCD can also store up to 120 COMSEC keys (traffic encryption key [TEK] and key encryption key [KEK]) or 280 transmission security keys (TSKs).

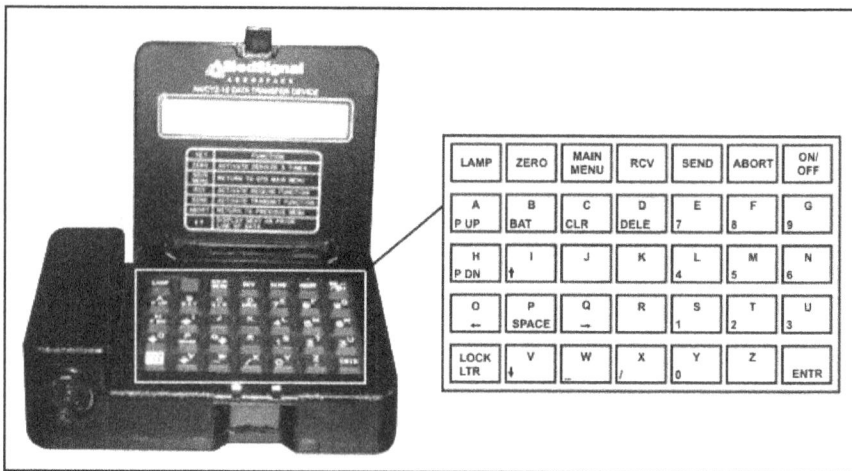

Figure 4-7. Automated net control device, AN/CYZ-10

4-34. The ANCD supports—

- **Memos**—receives, stores, and transfers up to four short memos, each six lines in length, with 22 characters per line.
- **Over-the-air rekeying (OTAR)**—supports both automatic keying and manual keying.
- **Broadcasts**—transmits SOI information from one location to another electronically.
- **Secure telephone units (STUs)**—allow COMSEC keys, FH data, and SOI information to be sent from one location to another.
- **Precision lightweight global positioning system receiver (PLGR)**—is capable of being loaded with the required operational key through the use of the ANCD.

COMPUTER SYSTEM, DIGITAL AN/PYQ-10

4-35. The AN/PYQ-10, simple key loader (SKL), was designed as a replacement for the AN/CYZ-10, ANCD. (Refer to Figure 4-8 for an example of the SKL.) A limited understanding of the Electronic Key Management System (EKMS) operating environment is helpful in understanding the operation of the SKL. The components of the EKMS include—

- **EKMS Tier 0.** The National Security Agency (NSA) central facility provides for production, management, and distribution of specialized electronic cryptographic key and associated materials.
- **EKMS Tier 1.** Facilities serve as focal points for the production, management, and distribution of service unique electronic cryptographic key and materials. Tier 1 facilities also provide an interface between the central facility and service EKMS Tier 2 elements, and facilitate interoperability for joint operations at the theater and strategic levels.
- **EKMS Tier 2.** Tier 2 or local communications security management software (LCMS) workstations perform generation, management, and distribution of electronic keying material. The LCMS workstation works in conjunction with the SKL to distribute electronic keying material to those networks with electronically keyed COMSEC equipment.
- **Automated communications engineering software (ACES) workstations.** The ACES workstation integrates cryptonet planning, electronic protection (EP) distribution, and SOI generation, management, and distribution. The ACES workstation works in conjunction with the SKL to automate cryptonet control operations for networks with electronically keyed COMSEC equipment.

- **EKMS Tier 3.** Tier 3 or the SKL device integrates the functions of COMSEC key management, control, distribution, EP management, SOI management, benign fill, and other specialized capabilities into one comprehensive mobile system. The SKL will interface with the ACES and LCMS workstations to receive its database information and then interface with end cryptographic units to upload the required keying material and information to those units.

4-36. The hardware platform that hosts the SKL software (including the Secure Library) is a vendor supplied ruggedized personal digital assistant device equipped with a KOV-21 Personal Computer Memory Card International Association card. The SKL is not equipped with a hard drive so all programs are stored in non-volatile flash memory.

4-37. The KOV-21 provides Type I encryption/decryption services and provides the secure interface between the host computer and interfacing devices. The SKL uses an embedded KOV-21 approach. As such, the NSA requires that a CIK be used to lock and unlock the KOV-21 information security card.

4-38. The CIK is a separate, removable, non-volatile memory device designed to protect internal SKL keys and data from physical compromise when the SKL is in an unattended, non-secured environment. When the CIK is removed from the SKL, the KOV-21 card cannot be unlocked. Therefore, access to the data is denied. The absence of the CIK prevents the use of SKL operations. (Refer to TM 11-7010-354-12&P for more information on the SKL.)

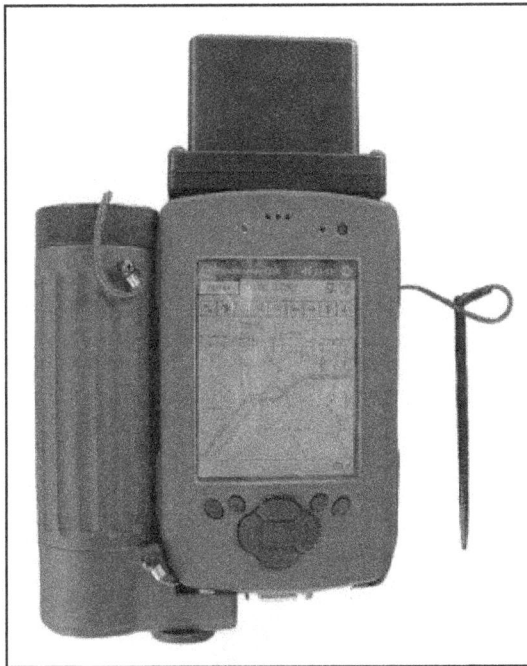

Figure 4-8. AN/PYQ-10 simple key loader

NAVIGATION SET, AN/PSN-11 PRECISION LIGHTWEIGHT GPS RECEIVER

4-39. The PLGR is a self-contained, handheld, five channel single frequency GPS receiver that provides accurate position, velocity, and timing data to individual Soldiers and integrated platform users. (Refer to Figure 4-9 for an example of the PLGR.) The PLGR computes accurate position coordinates elevation, speed, and time information from signals transmitted by the GPS satellites.

4-40. The PLGR selects satellites that are 10 degrees or more above the horizon (elevation angle) during initial acquisition. The PLGR requires a minimum of three satellites for location and four for elevation. It also utilizes precise timing from satellite to receiver to determine location and elevation and LOS to satellite receiver and COMSEC allow maximum accuracy. It can also be used with an external power source and an external antenna. Features of the PLGR include—

- Continuous tracking of up to five satellites.
- Course/Acquisition, precise, and encrypted P code capability.
- One handed operation.
- Backlit display and keyboard for night operation.
- Operates in all weather, day and night.
- Produces no signal that can reveal your position.
- Automatically tests itself during operation.
- Capability to store up to 999 waypoints.
- Capability to stores up to 15 routes with up to 25 legs per route.
- Sealed against dust and water to a depth of 1 meter (3.2 ft).
- Compatible with night vision goggles.

4-41. FH radios such as the SINCGARS depend on accurate time as part of the FH scheme. The PLGR supports SINCGARS in terms of precise time synchronization. GPS time is loaded into the SINCGARS from the PLGR. This data is loaded from connector J1 on the PLGR. The time figure of merit must be seven or less and have a SINCGARS connected. (For more information on the PLGR refer to TB 11-5825-291-10-2 and/or TM 11-5825-291-13.)

Figure 4-9. AN/PSN-11 precision lightweight GPS receiver

AN/PSN-13 (A) DEFENSE ADVANCE GPS RECEIVER

4-42. The defense advance global positioning system receiver (DAGR) was designed to replace the PLGR. The DAGR collects and processes the GPS satellite link one (L1) and link two (L2) signals to provide position, velocity, and timing information, as well as position reporting and navigation capabilities. The DAGR is primarily a handheld unit with a built-in integral antenna, but can be installed in a host platform (ground facilities, air, sea, and land vehicles) using an external power source and an external antenna. When the DAGR is used as a handheld unit it can also operate with an external L1/L2 antenna external power source. (Refer to Figure 4-10 for an example of the DAGR compared to a PLGR.)

Figure 4-10. AN/PSN-13 DAGR compared to a PLGR

4-43. Equipment capabilities and features of the DAGR include—

- Signal acquisition using up to 12 channels.
- Navigation using up to 10 channels.
- Accepts differential GPS signals.
- Backlit display and keypad for night operation.
- Operates in all weather, day or night.
- Produces no signals that can reveal your position.
- Automatically tests itself during power up.
- Operates on +9 to +32 volts direct current (VDC) external power.
- Performs area navigation functions, storing up to 999 waypoints.
- Stores up to 15 routes with up to 1000 legs for each route.
- Resists jamming.
- Resists spoofing when cryptographic keys are installed.

- Sealed against dust and water to a depth of 1 meter (3.2 ft) for 20 minutes.
- Interconnects with other electronic systems.
- Uses quick disconnect connectors and fasteners to allow easy unit replacement.
- Compatible with night vision goggles and does not trigger blooming.
- Uses internal compass to compute track and ground speed when moving at or below 0.5 meters (1.6 ft) per second.

DAGR and SINCGARS

4-44. The DAGR has a precise positioning service, HAVEQUICK, and SINCGARS page that is used to configure a DAGR communication port for a time synchronizing output from the DAGR (using external connectors J1 or J2) to another piece of equipment, such as a SINCGARS. (For more information on the DAGR refer to TB 11-5820-1172-10 and TM 11-5820-1172-13.)

VEHICULAR INTERCOMMUNICATIONS SYSTEM, AN/VIC-3

4-45. The VIS, AN/VIC-3 provides communications among crewmembers inside combat vehicles and externally over as many as six CNRs. (Refer to TM 11-5830-263-10 for operators' level information on the VIC-3) The VISs active noise reduction (ANR) capability offers significant improvements in speech intelligibility, aural protection, and vehicle crew performance. Figure 4-11 shows the VIS components.

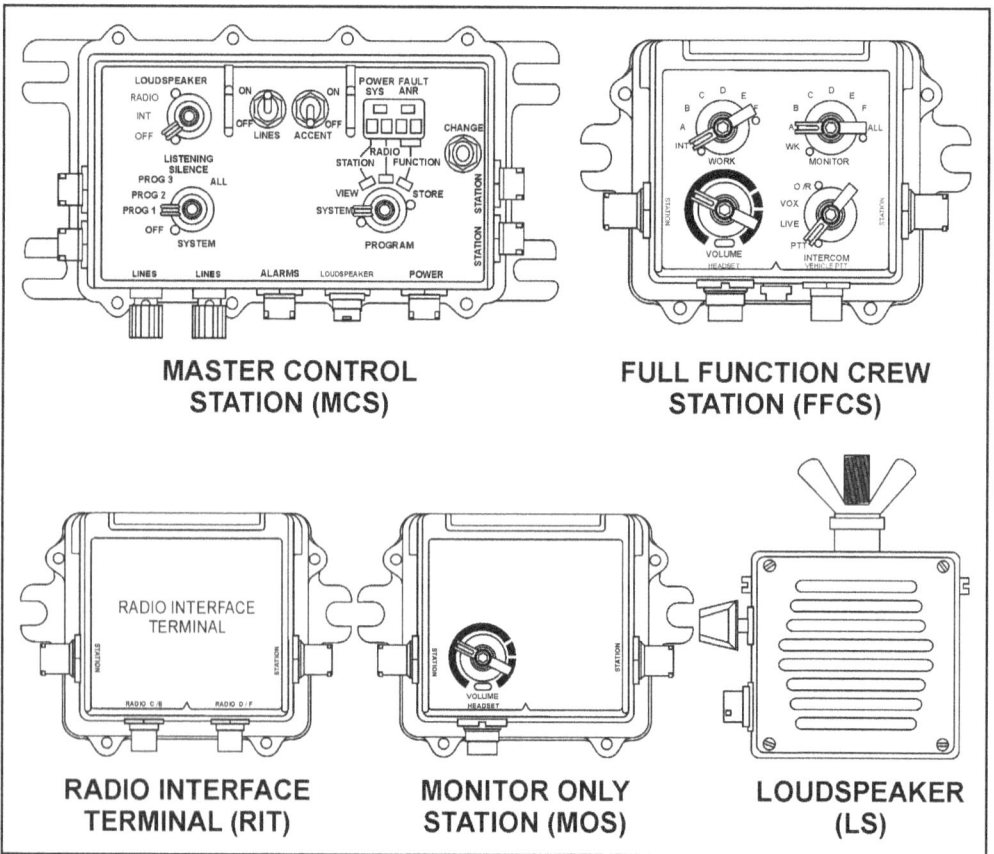

Figure 4-11. Vehicular intercommunications system, VIC-3 components

Master Control Station

4-46. The master control station (MCS) is the central node of the VIS. It connects directly to vehicle prime power, and provides the rest of the system with regulated power. The MCS provides connections for up to two radio transceivers, vehicle alarms, a loud speaker, and a pair of field wires (used to connect a field telephone, another MCS, GRA-39, SRCU, or an AN/VIC-1). The MCS performs a BIT routine on power-up, and continuous performance monitoring of the system.

4-47. The MCS contains built-in radio programming, providing control of radio access at all stations. The MCS allows the vehicle commander to enter five radio access operating modes using the system switch. Three of the modes are fully programmable, and when programmed, they contain specific rules that govern the radio transmit and radio receive access for each individual crewmember. Complete program rules are established on board at any time without external equipment.

Full Function Crew Station

4-48. The full function crew station is the interface between the VIS and the combat crewmember headset. It controls the headset volume and provides the user access to up to six onboard radios. The radio selections allow the user to communicate on any one radio while monitoring traffic on an additional radio or all radios. The full function crew station provides live (hot-mike) or voice operated keying facilities for hands free operation. An override facility provides an emergency position whereby the operator can force his intercom signal to all other crewmembers.

Monitor Only Station

4-49. The monitor only station provides a listen-only intercom capability for crewmembers in vehicles. All monitor only stations can operate through the vehicle slip rings. The monitor only station provides independent control of the headset volume.

Radio Interface Terminal

4-50. The radio interface terminal, which has no user adjustable controls, provides an interface for two additional radio transceivers, enabling the basic two radio systems to be expanded to six radios. The radio interface terminal design (two station ports and two dual radio ports) and the VIS ring architecture will accommodate radio placement above and below slip rings.

Loudspeaker

4-51. The system loudspeaker is connected to the MCS, broadcasting vehicle intercom or radio messages. A single switch mounted on the MCS controls loudspeaker power and message traffic.

Headsets

4-52. These headsets employ ANR and/or passive noise reduction to achieve noise reduction and enhance audibility. ANR is accomplished (when turned on) by electronic generation of noise canceling acoustic waves within each ear cup. Passive noise reduction is accomplished by soft conformal ear seals that are snug against the head and alleviate or reduce outside noise. All headsets are connected to the VIS system via a standard audio connector and quick-disconnect bailout connector to enable rapid disconnection from the system. There are five different types of headsets. Refer to Figure 4-12 for an illustration of the headsets.

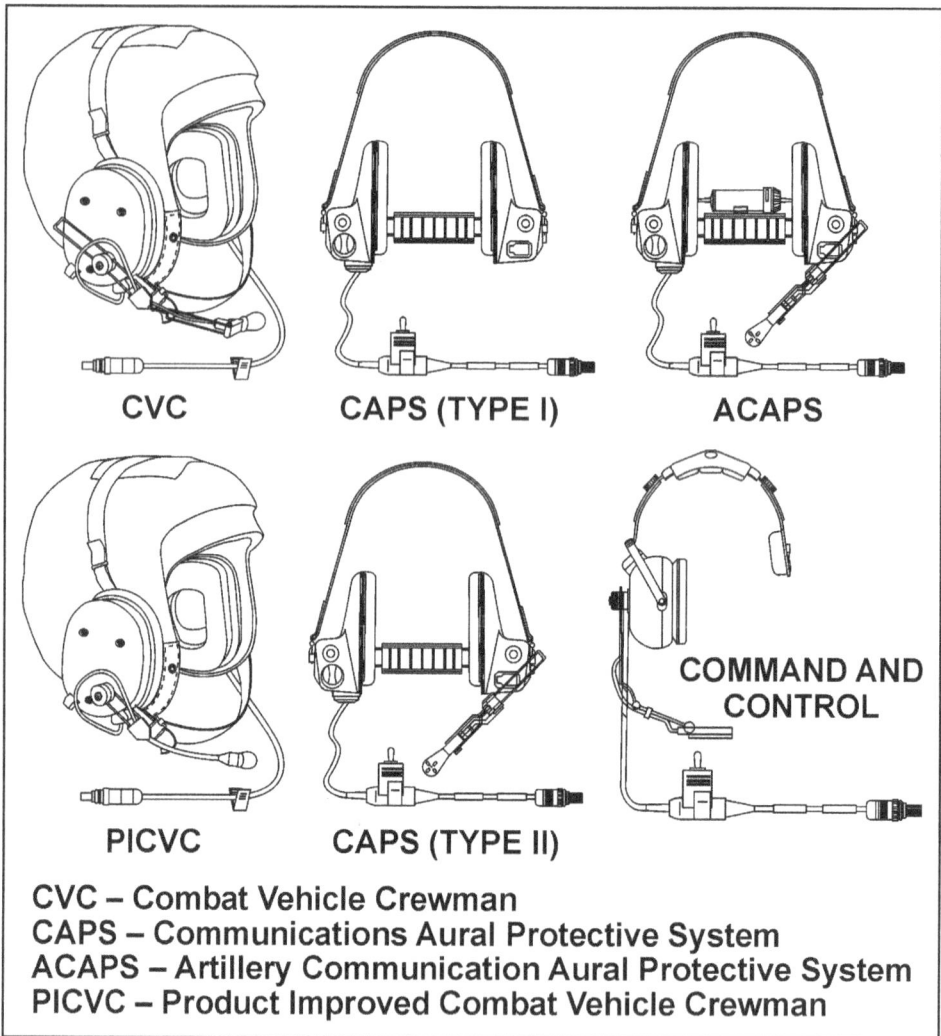

Figure 4-12. Vehicular intercommunications system, VIC-3 Headsets

HANDHELD REMOTE CONTROL RADIO DEVICE

4-53. The handheld remote control radio device (HRCRD), C-12493/U, is used with the manpack radio AN/PRC-119A/D/F and the dismount kits of vehicular radio configurations (AN/VRC-88A/D/F and AN/VRC-91A/D/F). (Refer to TM 11-5820-890-10-6 for more information on the operations of the HRCRD and SINCGARS.)

4-54. Figure 4-13 is an example of a HRCRD. The HRCRD enables the remote operator to control the following functions of the radio—

- Channel.
- RF power.
- Mode.
- COMSEC.

- Audio volume level.
- Back light.

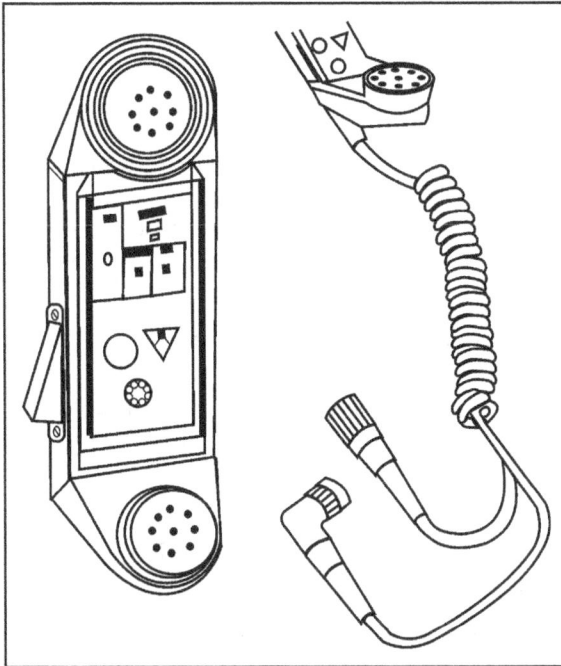

Figure 4-13. Handheld remote control radio device

SINGLE-CHANNEL GROUND AND AIRBORNE RADIO SYSTEM PLANNING

4-55. The initial operation plan (OPLAN) and the units SOP determine the type of net(s) needed. The network planner must answer the following questions—

- What type of information is passed: data, voice, or both?
- Does the unit require communications with users normally not in its network?
- Is the network a common-user or a designated membership network?
- Is wireless network extension needed to extend the network's range?

4-56. The unit G-6/S-6, assistant chief of staff, operations (G-3) and operations officer (S-3) work together to answer all these questions. Once these questions are answered, initial planning and coordination of the network can begin. Many of the items will become part of the units SOP. (Refer to Appendix A for information on FM networks.)

DATA NETS

4-57. The SINCGARS interfaces with several types of DTE, such as the secure digital net radio interference, TSEC/KY-90. SINCGARS also provides automatic control of the radio transmission when a data device is connected. It disables the voice circuit during data transmissions, preventing voice input from disrupting the data stream; disconnecting the data device during emergency situations overrides the disable feature. A single cable from the DTE to the radio or mounting adapter connects most DTEs.

SECURE DEVICES

4-58. The SINCGARS uses an internal COMSEC module. The encryption format is compatible with VHF/UHF wideband tactical secure voice system cryptographic equipment devices, provided they are loaded with the same TEK. SINCGARS uses the KY-57/58 (VINSON) for non-ICOM airborne radio systems.

4-59. The VINSON secure device has six preset positions: five for the TEK and one for a KEK. The TEK positions allow operation in five different secure nets. The KEK position allows changing or updating the TEK through OTAR. The ICOM secure module retains one TEK per preset hopset/net identifier (NET ID), and one KEK.

4-60. The variables are loaded and updated the same in both devices. The ANCD does the initial loading; variables can be updated by a second manual fill or by OTAR. In accordance with COMSEC regulations, only the TEK may be transmitted over the air. The KEK must still be physically loaded into either the VINSON or ICOM radio. Encryption variables are controlled through COMSEC channels and are accounted for per Army Regulation (AR) 380-40. (Refer to Appendix F for information on COMSEC compromise recovery procedures.)

4-61. Data input to the radio is interleaved into the radio's digital data stream. The VINSON or ICOM circuits encrypt the data before transmission. However, digital data may be encrypted before inputting the information into the radio. COMSEC variables must be common for the transmitting and receiving terminals; this is coordinated between the two units passing information.

SINGLE-CHANNEL GROUND AND AIRBORNE RADIO SYSTEM WIRELESS NETWORK EXTENSION STATION

4-62. Due to the limited number of SATCOM channels available in an AOR, there is a crucial need for single channel push to talk capability at the theater, corps, and division. Most of the SATCOM channels available in an AOR are controlled and assigned at the corps level and higher, and FM communications at corps, division, and brigade is used to provide SC communications on the move. FM wireless network extension is the most available means of addressing the crucial need for single channel push to talk capability at the theater, corps, and division. FM wireless network extension extends SC communications around obstacles and across increased distances to its subordinate units.

4-63. The commander (with the recommendation of the signal officer) decides the critical nets requiring wireless network extension support. Wireless network extension assets are primarily used to provide support for the following nets—

- C2.
- Administrative and logistic (A&L).
- Operations and intelligence (O&I).
- Fires.

Note. Refer to Appendix A for more information on FM networks.

4-64. The wireless network extension station operates on the command network to which it is subordinate, unless specifically tasked to operate on another net. The primary radio monitors the C2/O&I net; the secondary radio provides the wireless network extension link. Prior planning provides the wireless network extension station with the appropriate variables for the command net and wireless network extension net. The unit SOP should direct how the wireless network extension variables are assigned in accordance with possible alternatives.

4-65. SINCGARS can operate as either a secure or non-secure wireless network extension station. These radios automatically pass secure signals even if the wireless network extension radios are operating non-secure. However, the wireless network extension RTO cannot monitor the communications unless the secure devices are filled and in the cipher mode.

WIRELESS NETWORK EXTENSION PLANNING

4-66. Wireless network extension planning must be linked to the military decision making process to ensure success. During wireless network extension planning the S-6—

- Ensures the communications operations course of action (COA) is integrated into the maneuver COA.
- Plots primary and secondary wireless network extension locations on the COA sketch. Location selection must consider mission, enemy, terrain and weather, troops and support available, time available, civil considerations (METT-TC) analysis.
- Determines whether site collocation with another unit is required. Consideration must be given to security/sleep plan, logistics, and evacuation.
- Plans for contingency sites and establishes criteria, known to all concerned, that will initiate relocation/evacuation procedures.
- Develops reporting procedures to the establishing headquarters.
- Builds a wireless network extension team equipment list, and considers including the following communications equipment—
    - PLGR/DAGR.
    - ANCD/SLK.
    - Two OE-254 antennas with additional cables.
    - Any extra AN/PRC-119 radios (can be used as a backup RT).
    - Additional batteries.
- Establishes a pre-combat checklist and rehearses prior to deployment.

WIRELESS NETWORK EXTENSION MODES

4-67. The SINCGARS (ground) has built-in wireless network extension capability that requires the addition of a wireless network extension cable (CX-13298) for operations. SINCGARS can perform the wireless network extension function three ways. The network can be—

- Set up for SC to SC.
- Made of mixed modes (FH to SC or vice versa).
- Used in its full capability of FH to FH.

4-68. These options make wireless network extension flexible in operation. They also increase the prior coordination required before deployment. This ensures all users have access to the wireless network extension function.

Single-Channel to Single-Channel Operations

4-69. SC to SC operations require a 10 MHz separation between the frequencies (as shown in Figure 4-14, Wireless network extension operations). Physically moving antennas farther apart or lowering power output lessens the frequency separation. Table 4-3 shows the minimum antenna separation distance. The network NCS must monitor the wireless network extension station to ensure the command hopset is maintained. This ensures continuous communications for the unit.

Note. All RFs used should be obtained from unit SOIs which are coordinated with the unit electromagnetic spectrum manager. Units can not establish their own wireless network extension frequencies without electromagnetic spectrum manager coordination.

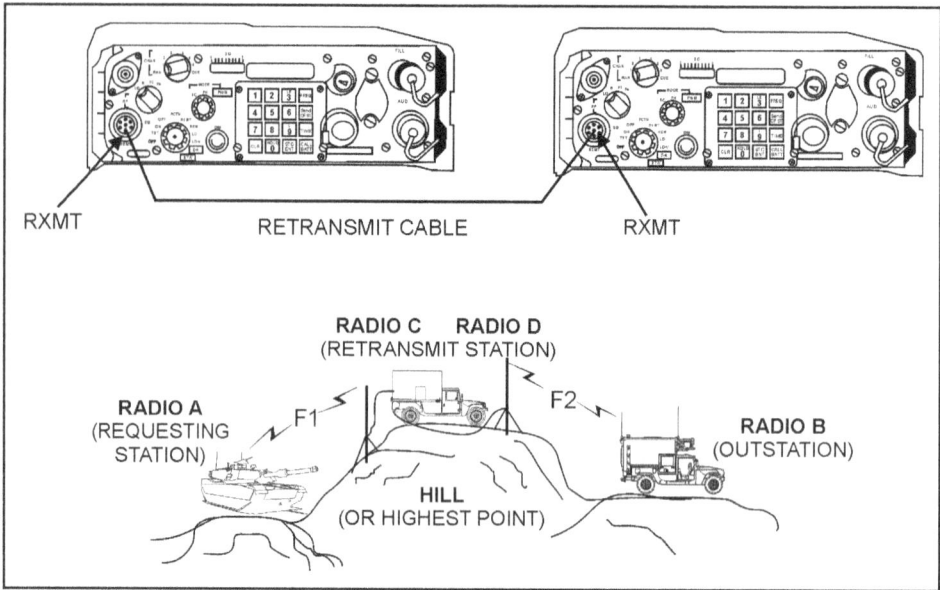

Figure 4-14. Wireless network extension operations

Table 4-3. Minimum antenna separation distance

Minimum Frequency Separation Required	High Power Separation	PA Power Separation
10 MHz	5 ft (1.5 meters)	5 ft (1.5 meters)
7 MHz	10 ft (3 meters)	60 ft (18.2 meters)
4 MHz	50 ft (15.2 meters)	150 ft (45.7 meters)
2 MHz	200 ft (60.9 meters)	400 ft (121.9 meters)
1 MHz	350 ft (106.6 meters)	800 ft (1463 meters)

Frequency Hopping to Single-Channel Operations

4-70. FH to SC operations is a simple mode to set up and operate with no requirement for frequency or physical separation. The SC frequency should not be part of the hopset resource used on the FH side of the wireless network extension. This method allows a SC radio user access to the FH net in an emergency situation. Continual access to the FH net using this method should be avoided to prevent lessening the ECCM capability of the SINCGARS.

Note. The wireless network extension station typically functions as the NCS during FH wireless network extension operations.

Frequency Hopping to Frequency Hopping Wireless Network Extension Operations

4-71. FH to FH wireless network extension operations allows for the wireless network extension of FH nets and is the simplest mode with no requirement for frequency or physical separation. FH wireless network extension operations will either be the traditional F1:F2 or F1:F1, depending upon the model of SINCGARS and mission. The SINCGARS ASIP provides the capability for F1:F1 operations.

4-72. F1:F2 operations require at least one of the NET IDs to be different (for example, NET ID F410 to NET ID F411). Any one, or a combination of NET IDs, may change. The preferred method is for the NET IDs, for each side of the wireless network extension, to be located within the same hopset. The wireless network extension station RTO functions like the network NCS for the outstation link. In this function, the RTO answers all cues, ERF, and authenticates net entry. The wireless network extension RTO must ensure the outstation RT is placed in the FH master mode; this ensures timing on this link is established and maintained.

4-73. F1:F1 operations allows for both NET IDs to be the same. This is important when operating in the tactical Internet. Wireless network extension is not an option in the packet mode for SIP and earlier SINCGARS, due to the critical timing associated with the packet mode. In a traditional F1:F1 wireless network extension, a member of the outstation could potentially have captured the net due to the relatively long delays encountered at the wireless network extension site; rendering the wireless network extension packet lost.

4-74. The ASIP system overcomes this problem by assigning each radio at the wireless network extension site as a dedicated receiver or transmitter. The ASIP shifts the incoming transmission by two hops in time and utilizes the same hopset on each leg of the wireless network extension (commonly called F1:F1). Therefore, packets are sent out the moment they are received without going through the process of interleaving (arranging data in a noncontiguous manner to increase performance) and deinterleaving. The shift in two hops is insignificant enough to affect the performance of the outstation and would make the wireless network extension site appear to be a part of one big net. (Refer to Appendix D for information on radio operations in unusual environments.)

SINGLE-CHANNEL GROUND AND AIRBORNE RADIO SYSTEM JAMMING AND ANTI-JAMMING

4-75. Jamming is the intentional transmission of signals that interrupts your ability to transmit and receive. If the radio signal is being jammed, the RTO will hear strong static, strange noise, random noise, no noise, or the net may be quiet with no signals heard. These signals depend upon the type of jamming signals and whether the radio net is operating in SC or FH mode. (Jamming and anti-jamming is addressed in detail in Chapter 11, Communications Techniques: EP.)

4-76. The simplest method the enemy can utilize to disrupt your communication is to transmit noise or audio signals onto a SC operating frequency, or on multiple FH frequencies during FH operation. If the enemy can generate enough power onto a unit's hopset, it is possible that communication capability will be disrupted or even stopped.

4-77. While SINCGARS is thought to be jam resistant due to its FH capability, in the event that SINCGARS is jammed, it may be necessary for you to take corrective actions. The action taken depends on the type of jamming or interference that is disrupting net communications as well as the authorized FH hopset that is available to the net.

4-78. When radio interference occurs, the RTO will determine if the interference is caused by jamming or equipment failure. To do this, the RTO will—
- Disconnect antenna; if noise continues, the radio may be faulty.
- Set the "function" FCTN switch to "squelch off" SQ OFF and listen for modulated noise.
- Look for a small signal strength indication on the RT front panel.

4-79. The following are corrective actions to take if jamming is indicated—
- Reposition or reorient antenna to eliminate interference.
- Notify supervisor of suspected jamming signals.
- Continue to operate.
- Work through jamming.
- Report interference and jamming to the NCS.

4-80. For those RT-1523F advanced system improvement program-enhanced (ASIP-E) pure nets, the NCS will make a net call in SC mode and instruct all net members to switch to FH mode 2 and continue to operate normally.

4-81. For those non-RT-1523F ASIP-E pure nets, the NCS will make a net call in SC mode and instruct all non-RT-1523F ASIP-E radios to switch to the backup SC secure frequency SC/CT. All RT-1523F ASIP-E radios will switch to FH mode 2. The NCS will operate the net in a FH mixed net operation utilizing a SINCGARS mixed-mode wireless network extension site/station to provide communications between the SC stations and the FH stations. Once the jamming source is neutralized, the NCS will instruct the net to switch back to FH mode 1.

Note. Operation of SINCGARS in the SC/CT mode should only be done when absolutely necessary.

AN/PRC-148 MULTIBAND INTER/INTRA TEAM RADIO

4-82. The multiband inter/intra team radio (MBITR) is used for company size nets depending on command guidance and mission requirements. It also has the capabilities of being used as a handheld radio to support the communications of a platoon, squad or team tactical environment for secure communications. It enables small unit leaders to adequately control the activities of subordinate elements. The MBITR can perform functions such as ground to air, ship to shore, SATCOM, civil military and multinational communications. The MBITR was first developed for use by SOF but many units throughout the Army (and other multinational governments) have seen an influx of the radio due to its value.

4-83. The MBITR radio set communicates with similar AM and FM radios to perform two-way communication. The AN/PRC-148 was built for frequency and waveform interoperability with legacy and newer systems (Joint Tactical Radio System [JTRS]). The radios concept was to ensure interoperability with virtually any common US military or commercial waveform operating in the 30–512 MHz frequency range with either FM or AM radio RF output, and with a user selectable power output from 0.1–5 watts. The AN/VRC-111 is the vehicular version of the MBITR. (Refer to Figure 4-15 for an example of the MBITR radio.)

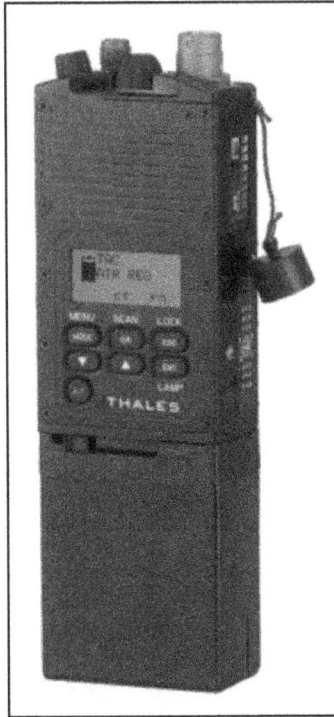

Figure 4-15. AN/PRC-148 MBITR radio

4-84. The MBITR is a portable, battery operated transceiver capable of providing both secure and non-secure communications. The MBITR operates in clear (analog) and secure (digital) voice and data. The basic radio is software upgradeable to add the following capabilities: SINCGARS, HAVEQUICK, ANDVT, and wireless network extension.

4-85. The MBITR has the following operating characteristics—

- Stores up to 100 preset channels organized in 10 groups with 16 channels each.
- SINCGARS voice and SIP data interoperable.
- HAVEQUICK I/II interoperable.
- ANDVT interoperable.
- Transmits voice in a whisper mode.
- Transfers configuration information to other MBITRs by means of a cloning cable.

4-86. The radio is tunable over a frequency range of 30–512 MHz, in either 5 or 6.25 kHz tuning steps, using 25.0 kHz channel bandwidth, 12.5 kHz when set for narrowband operation, and 5 kHz bandwidth when set for ANDVT. The radio automatically selects the correct tuning size.

4-87. The RT consists of a SC modulated carrier. The modulating source is analog or digitized voice and data signals at 12 (Federal Standard-1023) and 16 kbps (VINSON-compatible) in 25 kHz channel spacing. For emergency situations the radio circuitry is capable of receiving clear messages while set for secure mode operation.

AN/PRC-148 MBITR COMMUNICATIONS SECURITY

4-88. When operating in the secure mode, the radio disables the transmission of any tone squelch signals. Encryption key fill is accomplished through the audio/key fill connector. The urban MBITR has a standard

U-283/U six-pin connector that is fully compatible with the following key fill devices: KYK-13, KYX-15, KOI-18, and the AN/CYZ-10 (data transfer device [DTD]).

AN/PRC-148 MBITR and SINCGARS

4-89. SINCGARS operation is only available in those radios with the optional SINCGARS capability enabled. The following describes the transmission security (TRANSEC) capabilities of the MBITR with SINCGARS option—

- When operating in the SINCGARS mode, the available MBITR operating frequency range is 30–87.995 MHz.
- MBITR with SINCGARS functionality includes the operating modes of the basic MBITR radio and the following modes of operation—

    ■ SC clear FM analog voice operation, FM encrypted digital voice, and over-the-air FM transfer of encrypted digital data. The SC data mode implements the SINCGARS standard data mode and EDM.

    ■ FH PT digital voice operation, FH FM encrypted digital voice in 16 kbps continuously variable slope delta mode, and, using the SINCGARS and SINCGARS SIP waveforms, FH over-the-air FM transfer of encrypted digital data

AN/PRC-148 (MBITR) System Management

4-90. System management of the MBITR is the responsibility of the S-6 or communications section at all echelons. A Windows based personal computer (PC) radio programmer is provided to manage the quantity of radios. While all radio functions can be accomplished through the individual radio control panel if required, it would be very difficult to set up the radios for a battalion or larger force manually using radio front panel controls.

4-91. The PC programmer has a simple Windows "look and feel" interface that allows uploading and downloading information such as assigned frequency lists, waveform data, power level etc. to the radio. Once a radio is loaded with system information, it can be used to distribute this information (clone) to another MBITR. This cloning feature allows the S-6 system manager to distribute technical information down the tactical echelons to each individual radio in a command without fear of mistakes being made or data being corrupted.

AN/PRC-148 MBITR in Urban Operations

4-92. In small tactical units area coverage and distance extension has always been a problem. In urban operations communications inside buildings or over urban terrain has been a challenge. For these conditions the MBITR system provides a "back-to-back" (two radios) wireless network extension capability for both COMSEC and PT modes. Beside two radios, the only hardware required for wireless network extension is a small cable kit and some electronic filters. When configured for wireless network extension operations, a true digital repeater (digi-peater) is formed. Since the digits transmitted are merely being repeated by the radios they do not degrade signal quality and the radios do not have to have any COMSEC keys loaded in them.

AN/PRC-152 MULTIBAND HANDHELD RADIO

4-93. The AN/PRC-152 is a SC multiband handheld radio that has a JTRS architecture and software communications architecture. It also provides the optimal transition to JTRS technology. The AN/PRC-152 supports SINCGARS, HAVEQUICK II, VHF/UHF AM and FM. HAVEQUICK II and VHF/UHF AM and FM waveforms are ported versions of the preliminary JTRS library waveforms; validating the AN/PRC-152 JTRS architecture. Refer to Figure 4-16 for an example of an AN/PRC-152.

Figure 4-16. AN/PRC-152 multiband handheld radio

4-94. The AN/PRC-152 encryption device maximizes battery life in battery powered radios. It also supports all JTRS COMSEC and TRANSEC requirements as well as the ability to support numerous device compatibility modes: KY-57/VINSON, ANDVT/KYV-5, KG-84C, DS-101, and DS-102.

4-95. The AN/PRC-152 includes an embedded GPS receiver to display local position and to provide automatic position reporting for SA on the battlefield. The vehicular version of the AN/PRC-152 is the AN/VRC-110.

AN/PRC-152 VHF/UHF LINE OF SIGHT FIXED FREQUENCY PT

4-96. The AN/PRC-152 has the following VHF/UHF LOS operation frequency bands—

- **VHF low band** of 30.00000–89.99999 MHz.
- **VHF high band** of 90.00000–224.99999 MHz.
- **UHF band** of 225.00000–511.99999 MHz.

AN/PRC-152 VHF/UHF LINE OF SIGHT FIXED FREQUENCY CT

4-97. The AN/PRC-152 has following fixed frequency CT operation capabilities and limitations—

- **VINSON**—16 kbps data rate, 25 kHz wideband COMSEC (KY-57/58) mode for secure voice and data.
- **VINSON PT override**—alerts the RTO that a receiving transmission from an AN/PRC-152 in PT mode is being received.
- **KG-84C compatible**—(data only) supports secure data transmission in FM mode 30.00000–511.99999 MHz, and AM mode from 90.00000–511.99999 MHz. It is also used for UHF SATCOM operation.
- **TEK**—electronically loaded 128-bit transmission encrypted key used to secure voice and data communications.
- **COMSEC fill**—TEKs, TSKs, and KEKs can be filled from the following devices: ANCD, MX-18290, KYK13, and KYX-15.

Chapter 5

Ultra High Frequency Radios

This chapter addresses the UHF radios and systems that play a major role in network centric warfare such as the EPLRS, Blue Force Tracking (BFT), NTDR, JTIDS and MIDS.

FORCE XXI BATTLE COMMAND, BRIGADE AND BELOW

5-1. The Force XXI Battle Command, Brigade and Below (FBCB2) form the principal digital C2 system for the Army at brigade levels and below. It provides increased SA on the battlefield by automatically disseminating throughout the network timely friendly force locations, reported enemy locations, and graphics to visualize the commander's intent and scheme of maneuver. FBCB2 is a key component of the ABCS. Hardware and software are integrated into the various platforms at brigade and below, as well as at appropriate division and corps elements necessary to support brigade operations.

5-2. FBCB2 is a battle command information system designed for units at the tactical level. It is a system of computers, global positioning equipment, and communication systems that work together to provide near real time information to combat leaders. FBCB2 provides increased SA to the commander by depicting an accurate and automatic view of friendly forces, enemy forces, obstacles, and known battlefield hazards. FBCB2 provides enhanced SA to the lowest tactical level—the individual Soldier—and a seamless flow of C2 information across the battlefield.

5-3. FBCB2 supports OPCON through the transmission and receipt of orders, reports, and data. FBCB2 uses two forms of communications means: terrestrial and satellite. FBCB2 (terrestrial) uses EPLRS and FBCB2 (satellite) uses BFT. FBCB2 features the interconnection of platforms through EPLRS (terrestrial) and BFT (satellite) allowing the exchange of SA between the two systems. BFT systems share SA with EPLRS and EPLRS share SA with BFT systems and ABCS that use reachback tunnels found in regional operation centers.

ENHANCED POSITION LOCATION REPORTING SYSTEM

5-4. EPLRS supports the Army's transformation brigades, and is interoperable with the USAF, USMC and the USN. The EPLRS is currently employed in the C2 vehicle, battle command vehicle, Army Airborne Command and Control System (A2C2S), and TOC/tactical command post (TAC CP) platforms at the sustainment brigade and battalion level.

5-5. The EPLRS network is also the primary data communications system for the FBCB2, which is the data traffic backbone of the tactical Internet from brigade to lower echelons. The FBCB2 integrates with Army tactical C2 systems located within the brigade and battalion, and it provides real-time battlefield pictures at the strategic level. Using EPLRS communications and position location features, the FBCB2 integrates emerging and existing communications, weapon, and sensor systems to facilitate automated status, position, situation, and combat awareness reporting.

5-6. The EPLRS network provides the primary data and imagery communications transmission system. It is employed in the combat platforms of the commander, executive officer, first sergeant, platoon leaders, and platoon sergeants at the company and platoon level. The EPLRS is used as an alternate data communications link (host-to-host) between C2 platforms at the brigade and battalion level. It is the primary data communications link between battalion C2 platforms and company/platoon combat platforms. The EPLRS can be employed in wireless network extension platforms and configured to provide wireless network extension capability.

5-7. EPLRS is a wireless tactical communications system that automatically routes and delivers messages, enabling accurate and timely computer-to-computer communications on the battlefield. Using time division multiple access (TDMA), FH, and error correction coding technologies, the EPLRS provides the means for high-speed horizontal and vertical information distribution.

5-8. EPLRS radio sets are primarily used as jam-resistant, secure data radios that transmit and receive tactical data that typically includes—

- Operation orders (OPORDs).
- Fire support plans.
- Logistics reports.
- SA data.
- Cryptographic keys for radio sets.
- Configuration files for radio sets.
- E-mail.

5-9. The basis for EPLRS radio connectivity is the EPLRS needline. Each needline defines the operational relationship between the source and destination EPLRS units, without specifying which additional EPLRS units are part of the connection. The type of transmitted data, the mode of operation, and the data rate effects the planning distance between individual EPLRS units and the number of "hops," or relays, that can be included in an EPLRS link. Accurate planning and network configuration is critical to provide proper area coverage within the tactical environment. Refer to TB 11-5825-298-10-1, for more information on EPLRS and refer to TM 11-5825-298-13&P and FM 6-02.72 for more information on Enhanced Position Location Reporting System network manager (ENM).

ENHANCED POSITION LOCATION REPORTING SYSTEM

5-10. The EPLRS consists of an RT, an operator interface device (the user readout), an antenna, and a power source (refer to Figure 5-1). The radio set provides transmission relay functions that are transparent to the user.

5-11. The EPLRS radio set has the following characteristics and capabilities—

- Operates in the 420–450 MHz UHF frequency band.
- Provides secure, jam resistant digital communications and accurate position location capabilities.
- Uses TMDA, FH (512 times per second), and spread spectrum technology (eight frequencies between 420–450 MHz).
- Embedded COMSEC module, TRANSEC, and an adjustable power output provides secure communications with a LPI/D.
- BIT function that is activated at power turn on.
- Uses an omnidirectional dipole antenna capable of covering the 420–450 MHz frequency ranges.
- Provides wireless network extension functions that are transparent to the user. The maximum distance the EPLRS can cover is based on 3–10 km (1.8–6.2 miles) distance between each radio and the maximum number of relays in the link.
- Can handle up to 30 needlines. The maximum number of needlines available is dependent on the bps required for each needline.

5-12. There are four different configurations of the EPLRS—

- AN/PSQ-6 manpack radio set.
- AN/VSQ-2 surface vehicle radio set.
- ASQ-177C airborne radio set.
- AN/GRC-229 grid reference radio set.

Figure 5-1. Enhanced position location reporting system

5-13. The RF network consists of many EPLRS radio sets connected to host computers. This provides secure host-to-host data communications for the host computers.

5-14. The radio set uses a wide band direct sequence spread spectrum waveform, TDMA, FH, and embedded error correction encoding. These capabilities provide for secure, high speed data communications networked between ground units and between ground units and aircraft. Most of the radio sets attributes are programmable and this programmability lets the planner set up the best possible anti-jam performance and data rate for the unique operational environment and mission.

5-15. EPLRS has automatic relay capabilities to support BLOS coverage. These capabilities are automatically and continually adapted to the changing operational environment faced by a mobile communications system.

5-16. The radio set also supports position location and identification capabilities. Position location allows users to determine precisely where the user is. It is similar to, but independent of, the GPS. Using position location data from the radio sets, some hosts may have the capability to determine where other radio sets are and can perform navigation functions.

Enhanced Position Location Reporting System Needlines Functions

5-17. Needlines are also known as a logical channel number or permanent virtual circuit. There can be many needlines running on a radio set at one time, supporting the hosts' data communications needs. Needlines can be activated manually via the user readout or host, or automatically by the host. The radio set will automatically activate the needline if any data is received on the corresponding logical channel number. If the radio set is turned off or power is lost, active needlines will be automatically reactivated when the radio set is powered back on.

Types of Needlines

5-18. There are seven major types of needlines, each falling into the two major types of host-to-host services (broadcast and point-to-point)—

- **Point-to-point needlines** provide unequal data transfer capability for two endpoints' hosts. Either endpoint can have all the data transfer capability, or it can be split between them in various ratios. Data is transferred at user data rates from 1,200 bps each way, up to 56,000 bps all one way. An example of how a point-to-point needline works would be the same as one person talking to another person on a telephone.
- **Simplex (one-way) needlines** provide a single host the capability to send data to many hosts. For simplex needlines, data is transferred at user data rates from 160–3,840 bps. An example of how a simplex needline works would be the same effect as using a bullhorn to talk to many people at the same time who cannot talk back.

- **CSMA needlines** provide many hosts the capability to send data to each other. For CSMA needlines, data is transferred at user data rates from 150–487,760 bps (for the whole needline). The radio set ensures there are no other radio sets using the CSMA needline and then sends data from the host. When completed, another radio set will ensure no other radio sets are using the needline and then transmit, and so on. This protocol allows many endpoints' hosts (multiple access) to use the same CSMA needline to send data to one or more endpoints' hosts. An example of a CSMA needline would be like a group of people on a contention voice net, each speaking when they have something to say and no one else is speaking.

- **Multisource group (MSG) needlines** provide up to 16 hosts the capability to send data to many hosts. MSG needlines provide each source host guaranteed bandwidth without conflict, with user data rates from 37.5–485,760 bps. Data transferred from one source also goes to the other sources. If fewer sources are used, the sources can have more than $1/16^{th}$ of the data transfer capability. Each $1/16^{th}$ is called a share. For example, a source endpoint can be assigned to have $4/16^{ths}$ of the total MSG data transfer capability, with 12 other source endpoints each having $1/16^{th}$ of the total MSG data transfer capability. If there are unused shares, a radio set whose host load is larger than its assignment on the MSG needline will use these available shares. The more shares a radio set has, the more data transfer capability it has. The radio set also supports eight and four share MSG needlines that provide faster speed of service. An example of how an MSG needline works would be the same effect as up to 16 people with bullhorns talking, in a round robin fashion, to many people who cannot talk back. A MSG needline is similar to a CSMA needline, but each sender has a dedicated, guaranteed amount of time to talk (similar to many concurrent simplex needlines).

- **Low data rate duplex (two-way) needlines** provide radio-acknowledged, higher reliability, balanced data transfer between two hosts with data rates from 20–1920 bps each way. They provide equal data rates in both directions. This data transfer capability may be used by either or both endpoints. The endpoint radio sets will automatically ensure that the data is all delivered using radio set to radio set acknowledgement protocols. This needline type requires preplanning for the radio set to be able to use. An example of how a duplex needline works would be the same effect as talking to another person on a telephone.

- **Dynamically allocated permanent virtual circuit (DAP) needlines** are a special type of duplex needline. They have capabilities similar to those of duplex needlines (rates are 60–1920 bps), but DAP needlines are automatically set up and deleted on demand by the host, without any preplanning or NCS involvement. However, if the network resources are not available to support the data rate requested by the host, the needline rate is reduced to the highest rate available that the radio set can support.

- **High data rate (HDR) duplex needlines** have the same features as duplex needlines except that the data rates are higher, from 600–121,440 bps each way.

Enhanced Position Location Reporting System Communications Needlines Capabilities

5-19. An EPLRS radio set can support needlines as an endpoint, relay, or as both. A radio set can be a relay on some needlines, an endpoint on other needlines, and both an endpoint and a relay on other needlines, all at the same time. As an endpoint, a radio set can send and/or receive data to/from its host on a needline. A radio set that is only a relay (not an endpoint) cannot send or receive data to/from its host, and might not even have a host. For simplex, duplex, and DAP needlines, radio sets will automatically sign up as a relay if they have the resources available.

5-20. For point-to-point, CSMA, MSG, and HDR duplex needlines, a relay can only be endpoints on the needline, or they must be manually set up. When existing radio sets cannot support the EPLRS network relay needs, then dedicated relays are required.

5-21. There can be many host-to-host communications services running on a radio set at one time. There can be from one to thirty total needlines activated per radio set, depending upon the size of the needlines. If the maximum number is stored in the radio set, then another activated needline will cause the deletion of the oldest stored needline. There can be a maximum of eight activated CSMA, HDR duplex, MSG, and point-to-point needlines, total, per radio set.

5-22. A needline can use any of four waveform modes, 0–3. The higher the waveform mode number, the higher data rate capability the needline has, but the lower the needlines anti-jam capabilities. (For more information on EPLRS and system components refer to TM 11-5825-283-10.)

ENHANCED POSITION LOCATION REPORTING SYSTEM NETWORK MANAGER

5-23. The ENM equipment suite includes the following major components—

- **ENM software package (compact disk)**—ENM software program which includes installation program for loading ENM and EPLRS network planner onto ENM computer hard disk.
- **ENM computer**— consisting of a central processing unit and associated cabling; host computer platform for ENM software.
- **AN/VSQ-2D(V)1 surface vehicle radio set**—RT-1720DI/G, RT-1720EI/G, or RT-1720FI/G, with a user readout—also serves as the ENM radio set by connecting to the ENM computer.
- **Surface vehicle unit installation kit for SV-radio set**—includes platform, cables, user readout mount, and AS-3449/VSQ-1 antenna.
- **KOK-13 key**—generator key generation device for generating red and black cryptographic keys for network radio sets. (Not required for every ENM.)
- **KOI-18 tape reader**—device for inputting seed key data into KOK-13.
- **AN/CYZ-10 DTD (ANCD)**—key loading device for individually loading red keys into network radio sets; receives keys from KOK-13; physically connects to each radio set to accomplish loading.

Enhanced Position Location Reporting System Network Manager Characteristics and Capabilities

5-24. The ENM is a collection of software applications that run on a rugged host computer. The ENM software can run on Windows 2000 or Linux platforms and can be co-hosted with other applications as operational needs require. The ENM performs automated network management and control of the EPLRS network. The ENM assigns configuration parameters to radio set sets to allow them to perform their missions. The ENM manages the generation of cryptographic keys from a KOK-13 to load into the radio set.

5-25. The ENM application is installed on a rugged laptop computer and is used to configure a radio set and to plan, monitor, manage, and maintain an EPLRS network. Hosting ENM on a laptop computer also enables the ENM to be carried into the field for direct connection to a radio set for configuration and troubleshooting. The ENM computer physically connects to an EPLRS radio set called an ENM radio set directly via either Ethernet 802.3 or recommended standard-232 point-to-point protocol.

5-26. The ENM computer can also connect indirectly via a router using IP-over-Army Data Distribution System Interface Protocol. Refer to Figure 5-2 for an example of the EPLRS radio set and a host computer. The ENM vehicle is a high mobility multi-purpose wheeled vehicle that contains the ENM and other communications equipment.

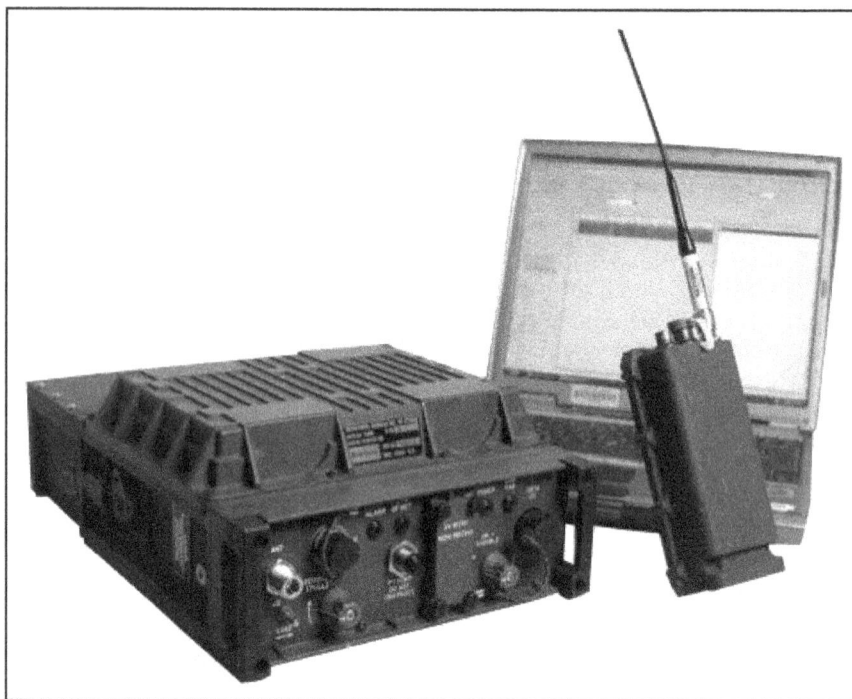

Figure 5-2. EPLRS radio set and host computer

5-27. The ENM was designed to plan and manage the EPLRS radio set network. The ENM can accommodate any size EPLRS radio set network. There are no restrictions on the number of radios that can be stored and managed by a single ENM. However, a maximum of 64 needlines can be assigned to any single radio set, so there are practical limitations to the size of the network.

5-28. The ENM's software application manages and controls the EPLRS network based on a deployment plan. ENM loads the radio sets with the configuration data needed to perform their missions. The ENM also generates the cryptographic keys for the radio sets. The ENM runs on a rugged laptop computer that connects to an assigned radio set.

5-29. The ENM operators set up, maintain, and manage the EPLRS network. There are two basic levels of ENM controlled by software login: network and monitor. Network ENMs monitor the status of network radio sets, configure radio sets over the air, initiate network timing, and perform other managerial tasks. Monitor ENMs have a lower level of access and only monitor the status of the network radio sets. They cannot perform the managerial task such as over-the-air reconfiguration of radio sets. Normally, one network ENM is made responsible for issuing the time master initiate command and distributing updated deployment plan files, if required. Other ENMs manage their own groups of assigned radio sets and coordinate with the time master ENM as required.

5-30. The EPLRS network is designed to maintain continuity of operations. If a specific ENM is disabled, control of the assets assigned to that ENM is automatically transferred to another ENM. Once an ENM is used to initiate the network, the existing network will continue to operate even if all ENMs were disabled. An ENM is not required for the radio sets to maintain the network and provide communications services to the hosts.

BLUE FORCE TRACKING

5-31. The BFT system is an L-band SATCOM tracking and communication system that provides the commander eyes on the friendly forces and the ability to send and receive text messages. BFT maintains SA of location and movement of friendly forces, sometimes termed "Blue Force," assets. BFT provides the Soldier with a globally responsive and tailorable capability to identify and track friendly forces in assigned areas of operations (in near real time), thereby augmenting and enhancing C2 at key levels of command.

5-32. The BFT contains computer hardware and software, interconnecting cables, L-band satellite transceiver, a PLGR, a mission data loader to transfer larger files, and an installation kit appropriate to the host vehicle type (if applicable).

5-33. The BFT computer console tracks friendly units carrying portable miniature transmitter devices. The transmitter devices are GPS-enabled, and send a signal via satellite detailing an individual or unit's location. Soldiers can program the transmitter devices to send location updates every five seconds. The transmitter devices are small enough to be carried in a Soldier's rucksack. Friendly forces appear as blue squares on the system's operator computer display. Units also have the capability to input enemy coordinate positions and obstacles on patrol routes. Enemy units appear as red squares, and obstacles as green squares. If units on the ground run into an enemy position, they can send that information to the system, and everyone who is connected on that network will be able to see the new data. The tracking system gives detailed information on friendly and enemy units up to a range of 5,000 miles. As long as the systems are connected through the satellite network, commanders can see the activities of their brigade and below-sized units.

5-34. The BFT supports a wide variety of joint missions and operations. BFT generates and distributes a common view of the operational environment at the tactical and operational levels, identifying and sharing that view with ground vehicles, rotary-wing aircraft, CPs, and Army and joint command centers.

NEAR TERM DIGITAL RADIO

5-35. The AN/VRC-108 is a LOS mobile packet radio network consisting of the NTDR and the network management terminal. (Refer to TM 11-5820-1171-12&P and TB 11-5820-1171-10 for more information on the NTDR.)

5-36. The NTDR (RT-1812) is a state-of-the-art, technology based digital radio. It is the primary data communications transmission system, linking the ABCS at the brigade and battalion echelons. The NTDR net provides a wireless wide-area network for Soldiers using ABCS host terminals located in TOCs. The NTDR wide-area network allows Soldiers to transmit information at HDR between TOCs, to support C2 data and imagery information flow. The NTDR (refer to Figure 5-3) net transceivers are typically employed in the following C2 platforms—

- The battle command vehicle in the First Digitized Division.
- C2 vehicles.
- Selected M1068 TOC and tactical platforms.
- UH-60 helicopters, equipped with the A2C2S.

Figure 5-3. Near term digital radio

5-37. The key features of the NTDR are that it—

- Operates in the UHF band (225–450 MHz) in discrete tuning steps of 0.625 MHz.
- Provides direct sequence spreading at a chip rate of 8.0 MHz, which enhances performance with respect to multiparous paths, jamming, and enemy interception.
- Provides nominal digital throughput at 288 kbps. Transmitted data is encrypted, protected with FEC and detection codes, and modulated onto an RF carrier. Received data is recovered following the same processes, but in reverse.
- Supports local area network (Ethernet) and serial (recommended standard-423 asynchronous and recommended standard-422 synchronous and asynchronous) interfaces.
- Includes a range of 10–20 km (6.2–12.4 miles).
- Incorporates a GPS receive capability that provides the military grid reference system position for the radio.

5-38. The brigade S-6, supporting the brigade OPLAN or OPORD, establishes NTDR networks to ensure successful network operation. This requires the establishment of separate cluster nets, and a backbone net to connect the clusters. A cluster may be formed by linking elements of a maneuver battalion together with the backbone linking the battalion clusters with the brigade TOC. Cluster heads form within the clusters to link the backbone, and to maintain connectivity. The NTDR has a self-organizing networking capability that provides highly mobile operations. End-to-end routing within the NTDR net structure is IP based.

TACTICAL DIGITAL INFORMATION LINK-JOINT TERMINALS

5-39. Tactical digital information link-joint (TADIL-J) is an approved data link used to exchange real-time information (NATO Link 16 is the near equivalent of TADIL-J). The TADIL-J is the protocol approved for joint (US only) air and missile defense surveillance and battle management. The TADIL-J is a communications, navigation, and identification system that supports information exchange between tactical communications systems. It is a secure, FH, jam-resistant, high capacity link, and uses the JTIDS or MIDS communications data terminal for both voice and data exchange.

5-40. JTIDS/MIDS operates on the principal of time TDMA, wherein time slots are allocated among participant JTIDS units for the transmission of data. This eliminates the requirement for an NCS by providing a nodeless communications architecture.

5-41. Army TADIL-J terminals are the JTIDS Class 2M and the MIDS low volume terminal (LVT)-2. Although other services' JTIDS and MIDS terminals exchange data and voice, Army JTIDS class 2M and MIDS LVT-2 terminals have no voice capability.

5-42. TADIL-J networks participants include—

- Joint Land Attack Cruise Missile Defense Elevated Netted Sensor System (JLENS).
- F/A-18.
- Airborne Warning and Control System (AWACS).
- E-2C Hawkeye aircraft.
- Tactical Air Operations Module (TAOM).
- Short-range air defense (SHORAD).
- Aegis ships.
- Medium Extended Air Defense System (MEADS).
- Patriot.
- Air Operations Center.
- Theater High Altitude Air Defense (THAAD).
- Air and Missile Defense Command.
- Joint tactical ground station (JTAGS).

TACTICAL DIGITAL INFORMATION LINK-JOINT TERMINALS AND ENHANCED POSITIONING LOCATION REPORTING SYSTEM NETWORKS

5-43. EPLRS is the primary data distribution system for forward area air defense C2 weapon systems. The typical SHORAD battalion use EPLRS to establish a data network that interconnects the Airspace Command and Control, Air Battle Management Operations Center, C2 nodes, platoon and section headquarters, and individual weapons systems. It passes the air picture and weapons control orders down, and then sends weapons systems status back up through the system. The extended air picture received from air and missile defense units, and E-3A Sentry/AWACS systems, are fused with the air picture received from the AN/MPQ-64, Sentinel, filtered at the forward area air defense C2 node for specific geographical areas of interest, and broadcast to all subscribers.

JOINT TACTICAL INFORMATION DISTRIBUTION SYSTEM

5-44. JTIDS is a UHF terminal that operates in the 960–1215 MHz frequency band. It uses the Department of Defense's (DOD's) primary tactical data link to provide secure, jam-resistant, high-capacity interoperable voice and data communications for tactical platforms and weapon systems. Using TADIL-J and the Interim JTIDS message specification, the Army JTIDS allows air defense artillery (ADA) units to exchange mission essential data in near real-time, with other Army joint communications organizations performing joint an AOR air and missile defense.

5-45. Army JTIDS supports joint interoperability and attainment of dominant SA, through integration of high throughput Link 16 messages, standard and waveform. Integrated in Army AOR air and missile defense weapons systems, Army JTIDS complements land force and joint force objectives for airspace control, by rapidly and securely supporting the exchange of surveillance, identification, unit status, and engagement information in both benign and electronic warfare (EW) conditions.

5-46. Host platforms for Army JTIDS/MIDS include—

- Forward area air defense command, control, and intelligence.
- Patriot power projection for Army command, control and communications.
- JLENS.
- THAAD.
- MEADS.
- JTAGS.
- Air and Missile Defense Planning and Control System at ADA brigades and Army air and missile defense commands.

5-47. The Army currently uses the JTIDS and/or MIDS at several operational levels as the medium to defense broadcast, and receive an enhanced joint air picture. An in-theater joint data net will provide shared joint C2 data and targeting information. Sources of the joint data net include—

- E-3A Sentry/AWACS.
- Control and reporting center.
- Intelligence platforms.
- E-2C Hawkeye aircraft.
- Aegis ships.
- Fighter aircrafts.
- USMC TAOM.
- Air defense and airspace management cell.
- ADA brigades.
- SHORAD.
- Patriot.
- THAAD.
- JTAGS.

5-48. The Army JTIDS system is comprised of the Class 2M terminal, the JTIDS terminal controller, and the JTIDS antenna. Figure 5-4 is an example of the JTIDS Class 2M radio, AN/GSQ-240 I.

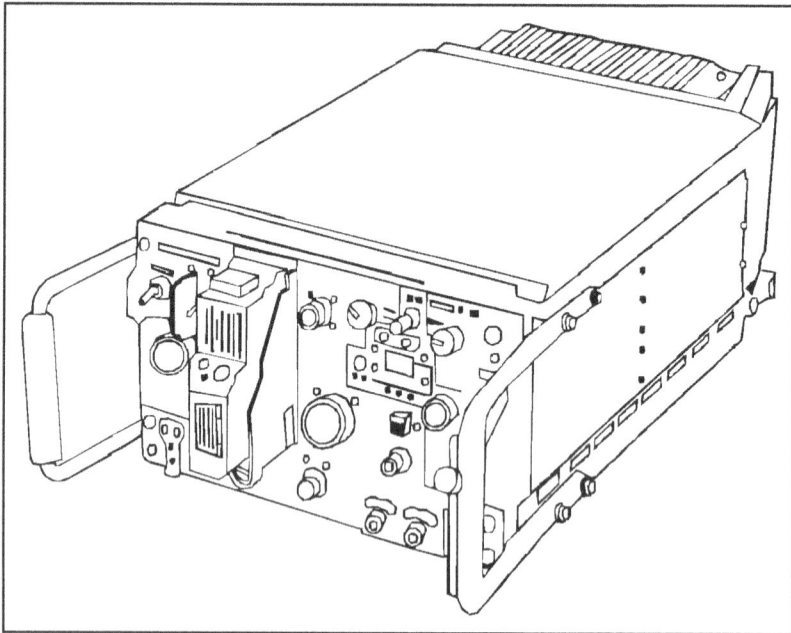

Figure 5-4. JTIDS class 2M, AN/GSQ-240 I radio set

MULTIFUNCTIONAL INFORMATION DISTRIBUTION SYSTEM

5-49. MIDS is a communications, navigation, and identification system intended to exchange C2 data information among various C2 and weapons platforms, to enhance varied missions of each Service. MIDS is the follow-on to JTIDS terminals, providing improvements over the Class 2 family of terminals. MIDS is smaller and lighter than its predecessor and can be installed in platforms that are limited in space and weight. MIDS-equipped platforms are fully compatible with LINK 16 participants.

5-50. Army MIDS consists of a MIDS LVT-2, a terminal controller, and an antenna. Figure 5-5 shows the Army MIDS LVT-2, AN/USQ-140. The Army MIDS provides jam-resistant, near real-time, high digital data throughput communications, position location reporting, navigation, and identification capabilities to host platforms.

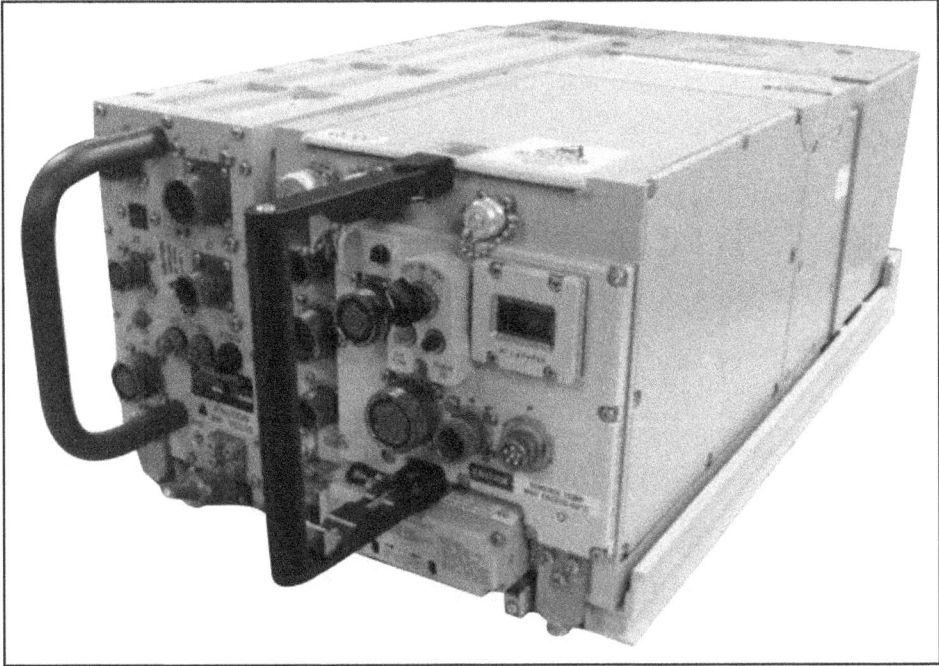

Figure 5-5. Army MIDS LVT-2, AN/USQ-140

This page intentionally left blank.

Chapter 6

Single-Channel Tactical Satellite

This chapter addresses the Army SC TACSAT planning and employment. It also addresses SC ground terminals, the AN/PSC-5, AN/PRC 117F and Army conventional forces.

SINGLE-CHANNEL TACTICAL SATELLITE INTRODUCTION

6-1. The Army uses SC TACSAT to provide long-haul, worldwide communications coverage to support critical C2 communications to ground and mobile operating forces. SC TACSAT provides the ability to support a small number of burst transmissions per day for SOF, Ranger units, atomic demolition teams, and long range surveillance units engaged in sensitive missions over extended distances and varied terrain. It also provides secure voice communications for C2 for the Special Operations Command, airborne, air assault, light infantry divisions, and light infantry brigades.

6-2. All Army SC TACSAT terminals provide half duplex operations. The radios provide the capability of transmitting data rates of up to 56 kbps on 25 kHz (wideband) channels and 9.6 kbps on 5 kHz (narrowband) channels. Due to the limited resources available on the UHF satellites and the increasing requirements for access by Army and all Service users, the Joint Chiefs of Staff (JCS) mandated the use of DAMA. This allows more access to the satellites through the automated sharing of the channel but reduces the data rates provided to the users. Therefore, the normal access is limited to 2400 bps, providing voice using ANDVT and data. The improvement of the voice encoder (VOCODER) in the radios using MELP vastly improves the voice quality and clarity at 2400 bps to that found using VINSON encryption at 16 kbps.

6-3. The JCS and SC TACSAT community realized there were problems previously experienced with the implementation of DAMA. MELP was not originally available and the users of voice found that narrowband did not provide what they needed to support their missions. In addition, DAMA was hard for operators to use and access could be preempted, causing the loss of communications during important missions. Most importantly, the satellites being used are failing due to surpassing their life cycle and the follow on system (both satellites and terminals), Mobile User Objective System, has been delayed.

6-4. The MIL-STDs governing the use of UHF were improved and implemented a higher data throughput into the sharing of channels. This is known as the integrated waveform. Implementation of integrated waveform is projected to take place in 2008. This will provide an improvement of up to four times the accesses seen in DAMA on a 25 kHz channel. Radios will be required to have the MELP VOCODER, providing the clear voice necessary for successful operations. Data rates on the channels can be changed on demand for those that need to send large data files in a short period. Integrated waveform will be implemented in two phases.

6-5. The first phase will allow for net communications, preplanned to support operations. Phase II will allow for ad hoc communications, supporting point-to-point calls and nets that were not preplanned as the mission dictates. The human machine interface will be simplified to allow for ease of operation. The prevalent manpack systems (AN/PSC-5C, AN/PSC-5D, and AN/PRC-117F along with the airborne system AN/ARC-231) will be the first to implement the integrated waveform.

SINGLE-CHANNEL TACTICAL SATELLITE PLANNING CONSIDERATIONS

6-6. The SC TACSAT mission provides worldwide tactical communications such as en route contingency communications, in-theater communications, intelligence broadcast, and CNR range extension. SC TACSAT radios link TOCs to all echelons, and include the long range surveillance units and Army SOF units, which can operate hundreds of miles from main forces.

6-7. Army SC TACSAT operates in the UHF band, and is available in manpack and vehicle versions. The radios' lightness, availability, and ease of use make them valuable for mobile and covert operations spanning full spectrum operations.

6-8. Commanders distribute terminals based on the mission and their preferences for communications. Commanders can use the terminal based on their vision of the battle scenario; flexibility and mobility are an inherent part of this architecture. Members of the Warfighter Network can be located anywhere within an AOR given the extent of the satellite footprint.

6-9. Unlike most communications systems SC TACSAT has no planning range. The capability to communicate depends on the location of the satellite for LOS. The channelization of each satellite is standardized providing flexibility and interoperability in normal operations. Given a contingency mission, the controlling authority can change the geosynchronous position of the satellite and improve the footprint as required. Additionally, SC TACSAT will not directly interfere with other combat net communications systems due to the frequency bands in which it operates.

DAMA NETWORKS

6-10. DAMA is a technique which matches user demands to available satellite time. Satellite channels are grouped together as a bulk asset, and DAMA assigns users variable time slots that match the RTOs information transmission requirements. The RTO does not notice a difference because the RTO appears to have exclusive use of the channel. The increase in nets or radio users available by using DAMA depends on the type of users. DAMA is most effective where there are many users operating at low to moderate duty cycles. This describes many tactical nets; therefore, DAMA is particularly effective with SC TACSAT systems.

6-11. DAMA efficiency also depends on how the system is formatted which is how the access is controlled. The greatest user increase is obtained through unlimited access. This format sets up channel use on a first-come-first-serve basis. Other types of formats are prioritized cueing access and minimum percentage access. The prioritization technique is suitable for command type nets, while the minimum percentage is suitable for support/logistic nets. Regardless of format, DAMA generally increases satellite capability by 4–20 times over normal dedicated channel operation.

SINGLE-CHANNEL ULTRA HIGH FREQUENCY AND EXTREMELY HIGH FREQUENCY TERMINALS

6-12. The following paragraphs address SC UHF and extremely high frequency (EHF) ground terminal radios.

LST-5B, 5C AND 5D

6-13. The LST-5B and LST-5C (refer to Figure 6-1) are SC TACSATs that operate in either a manpack, vehicular, shipboard, or airborne configuration. They are capable of operation by remote control via dedicated hardware, or PC-based software through an X-mode connector. Both radios modulate in AM and FM voice, cipher, data, and beacon. They use the frequency range of 225–399.995 MHz with channel spacing of 5 kHz and 25 kHz.

Figure 6-1. LST-5

6-14. The LST-5D has the added capability of DAMA, features embedded encryption devices for voice and data communications, as well as the channel capacity increases made possible through DAMA channel management. (Refer to FM 6-02.90 for more information on tactics, techniques, and procedures [TTP] for UHF TACSAT and DAMA operations.)

SINGLE-CHANNEL ANTI-JAM MAN PORTABLE TERMINAL, AN/PSC-11

6-15. The AN/PSC-11 single-channel anti-jam man portable (SCAMP) terminal is a man packable system that is packaged for storage or transport in two transit cases. The SCAMP consists of a RT, an interface unit that encrypts and decrypts the voice and data by using COMSEC keys, a handheld control device (30 key keypad), and a handset. (There is additional associated equipment that is not provided with all terminals.)

6-16. The AN/PSC-11 terminal interfaces with the military strategic and tactical relay system to provide secure, survivable voice and data communications via a low data rate payload. It can operate over EHF packages on fleet satellite and UHF follow-on systems. The AN/PSC-11 terminal operates in either point-to-point or broadcast modes, and provides voice and data service at a maximum data rate of 2,400 bps. The terminal can interface in the data mode with CNRs and PCs to provide range extension for conventional units and SOF. The AN/PSC-11 terminal has the following characteristics and capabilities—

- **Throughput**: 24 kbps (voice or data).
- **Modes of operation**: point-to-point or broadcast.
- **Frequency**: uplink, 43.5 to 45.5 gigahertz (GHz) Q Band with 2 GHz bandwidth.
- **Security**: embedded COMSEC.

6-17. The AN/PSC-11 terminal (refer to Figure 6-2) can interface with a variety of Army user communications systems via the four baseband data ports. The satellite link is transparent to the user communications system. The baseband equipment/systems do not control the satellite access of the

AN/PSC-11 terminal. In all cases, the operator must first establish the satellite path via the AN/PSC-11. Once the satellite path is operational, the baseband service can then be established. (Refer to TM 11-5820-1157-10 for more information on the SCAMP.)

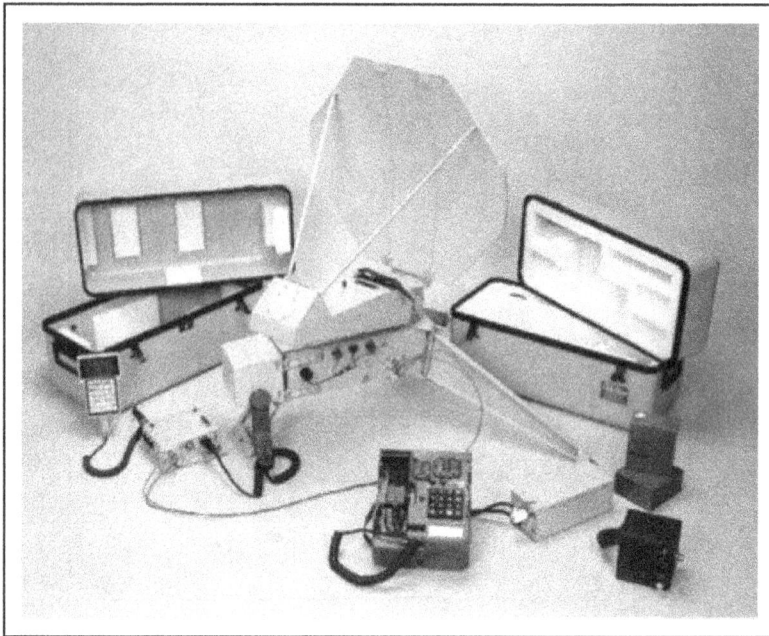

Figure 6-2. AN/PSN-11 SCAMP

AN/PSN-11 AND COMBAT NET RADIOS

6-18. The AN/PSC-11 terminal supports the SINCGARS SIP and ASIP radios, providing range extension to CNR users. The SINCGARS RT operates in the data mode only with the AN/PSC-11. Figure 6-3 shows the two AN/PSC-11/CNR configurations. With SINCGARS, the AN/PSC-11 operates in a full duplex, point-to-point configuration that supports user baseband equipment, such as the STU III and all utilized data systems. Additionally, the AN/PSC-11 provides range extension to the SINCGARS. The AN/PSC-11 can provide range extension to either a network or one SINCGARS. Connectivity via the red port or a black port (with an external cryptographic device such as the KG-84/KIV-7) provides encryption.

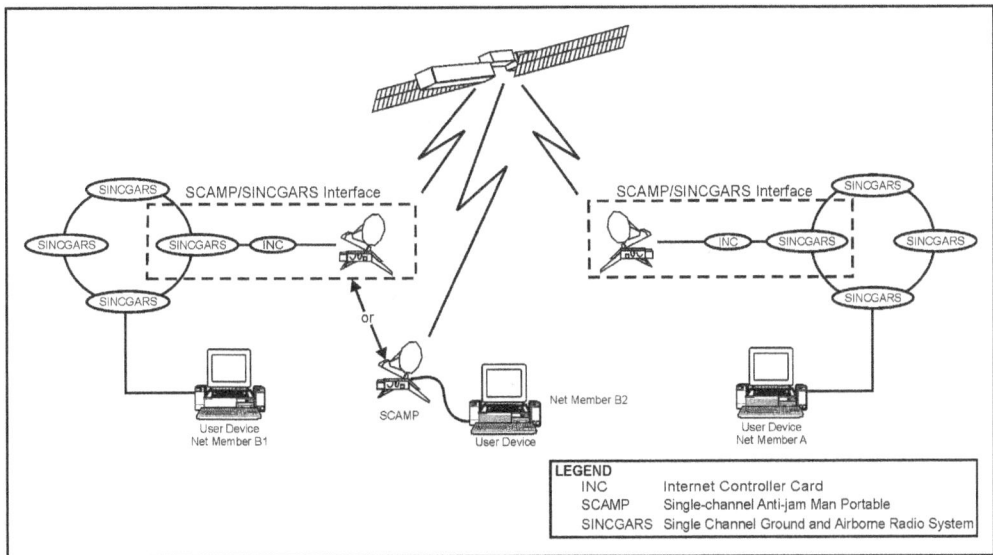

Figure 6-3. SCAMP/CNR configurations

AN/PSC-5 RADIO SET (SPITFIRE)

6-19. The AN/PSC-5 was built to replace the AN/PSC-3. Refer to Figure 6-4 for an example of an AN/PSC-5, Spitfire. The Spitfire operates in the following PT LOS modes with the following characteristics and capabilities—

- **Frequency bands** of:
    - 30.000–87.995 MHz.
    - 108.000–129.995 MHz.
    - 130.000–148.995 MHz.
    - 156.000–173.995 MHz.
    - 225.000–399.995 MHz.
- **Modulation** to include:
    - **AM**—60 to 90 percent at 1 kHz AM for PT and CT LOS voice modulation; 50 percent minimum for beacon mode.
    - **FM**—±5.6 kHz deviation at 1 kHz FM for PT and CT LOS voice modulation. The FM beacon modulation has a ±4 kHz nominal frequency deviation.
    - **FM**—frequency shift key (FSK) modulation rate of 16 kbps PT and CT voice and data. Used in LOS and SATCOM modes.
- **Channel spacing**: 5 kHz.
- **Squelch**: Operator adjustable S/N ratio squelch. From 10dB signal, noise and distortion (SINAD) at minimum squelched condition to at least 16 dB SINAD at maximum.
- **Half duplex operation**.
- **PT**: transmitted voice or data is not encrypted.
- **CT**: When a cipher-text voice message is received or transmitted (mode switch in CT), a single beep will be heard in the handset at the beginning of the reception or transmission.
- **Noise figure LOS**: 10 dB nominal.
- **Six presets**.

- **Frequency scanning:** capable of scanning five presets in LOS PT voice and CT (VINSON) voice.

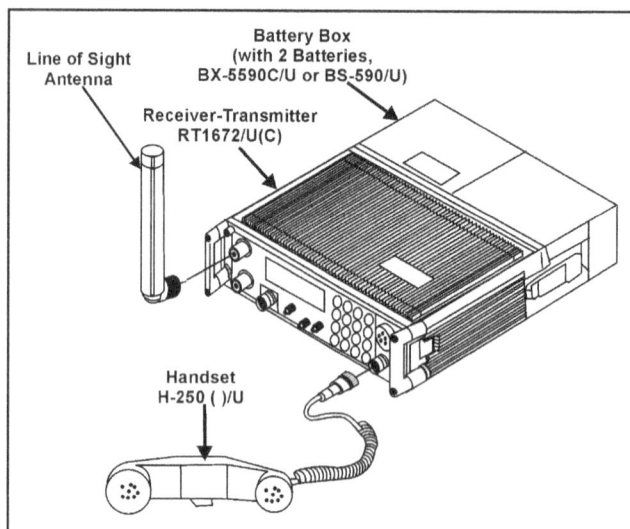

Figure 6-4. AN/PSC-5 radio set, Spitfire

6-20. The Spitfire can scan up to five LOS or dedicated SATCOM radio voice operation nets. Scanning combinations of CT (VINSON) and PT nets is allowed in voice mode only.

6-21. The Spitfire operates in the following SATCOM modes with these characteristics and capabilities—

- **Frequency band**: UHF band 225.000 MHz to 399.995 MHz.
- **Modulation** to include:
    - **AM**—60 to 90 percent at 1 kHz AM for PT and CT LOS voice modulation; 50 percent minimum for beacon mode.
    - **FM**—±5.6 kHz deviation at 1 kHz FM for PT and CT LOS voice modulation. The FM beacon modulation has a ±4 kHz nominal frequency deviation.
    - **FM**—FSK rate of 16 kbps PT and CT voice and data. Used in LOS and SATCOM modes.
    - **SBPSK**—modulation rate of 1200, 2400, and 9600 bps. Used in SATCOM mode.
- **Channel spacing**: 5 kHz and 25 kHz.
- **Squelch**: Operator adjustable S/N ratio squelch. From 10dB SINAD at minimum squelched condition to at least 16 dB SINAD at maximum.
- **Half duplex operation**.
- **PT**: transmitted voice or data is not encrypted.
- **CT**: When a cipher-text voice message is received or transmitted (mode switch in CT), a single beep will be heard in the handset at the beginning of the reception or transmission.
- **Noise figure SATCOM**: less than 4 dB (240–270 MHz).
- **Six presets**.

6-22. The Spitfire operates in the following DAMA modes with the following capabilities and limitations—

- **Frequency band**: UHF band 225.000–399.995 MHz.
- **Modulation** to include:
  - **Shaped offset quadrature phase shift keying (PSK)**—modulation rate of 600, 800, 1200, 2400, and 3000 bps used in 5 kHz DAMA mode.
  - **Binary PSK**—modulation rate of 19.2k and 9600 symbols per second used in 25 kHz DAMA mode.
  - **Differentially encoded quadrature PSK**—modulation rate of 32,000 symbols per second used in 25 kHz DAMA mode.
- **Channel spacing**: 5 kHz and 25 kHz, IAW MIL-STD 188-181, 188-182A, and 188-183.
- **Half duplex operation**.
- **VINSON**: 16 kbps data rate, 25 kHz COMSEC (KY-57/58) mode for secure voice and data.
- **KG-84** compatible modes 3 and 4 (data only).
- **ANDVT**—2400 bps mode for secure voice and data.
- **Six sets DAMA** (including 20 "sub-presets" each for 5 kHz service setup, 5 kHz message setup, and 25 kHz service setup).

Spitfire Wireless Network Extension Capabilities

6-23. The Spitfire provides range extension for both SINCGARS and Spitfire radios. A Spitfire-to-Spitfire wireless network extension is used when the network spans two satellite footprints. The actual terminals used for wireless network extension are set up in the PT mode, a W-5 cable is used between the two radios with SATCOM antennas connected, and the set up does not allow for an eavesdrop capability at the wireless network extension site.

Note. Do not attach handsets or speakers to Spitfire terminals in the wireless network extension configuration. If connected they will produce a non-secure beep broadcast and NSA mandates secure, encrypted transmissions.

6-24. The Spitfire terminals may be set up in the wireless network extension mode with the LOS antennas connected, but this is not recommended. A SINCGARS wireless network extension configuration is recommended for this communications requirement.

6-25. The abbreviated wireless network extension mode for SINCGARS requires one Spitfire to be set up with a SINCGARS at the wireless network extension site. Again, the Spitfire must be in PT mode to accomplish the wireless network extension, or eavesdropping may take place at the SINCGARS terminal. The SINCGARS operates in 25 kHz increments, the same as the LOS mode for the Spitfire. Both SATCOM and DAMA, 5 kHz channels, must be requested for the Spitfire to accomplish the communications link. The Spitfire set up at the distant end will be in the CT mode. It will then encrypt/decrypt transmissions using the COMSEC employed by the SINCGARS.

6-26. Use the AN/PSC-5 for BLOS wireless network extension of SINCGARS nets. Each net requires a SINCGARS and AN/PSC-5 terminal connected for wireless network extension. Figure 6-5 is an example of a SINCGARS range extension with AN/PSC-5 (this configuration can also be modified for SINCGARS to Spitfire communications).

6-27. In the PT mode, the wireless network extension AN/PSC-5 cannot monitor the network or send messages; only the SINCGARS terminal can do this. Additionally, satellite channels must be in 25 kHz increments for both SATCOM and DAMA. Once this configuration is complete, wireless network extension occurs as if it were a SINCGARS-to-SINCGARS wireless network extension site. The major difference is that the network at each end has BLOS capability.

6-28. Other available wireless network extension capabilities include DAMA-to-DAMA, DAMA-to-SATCOM, SATCOM-to-LOS, and DAMA-to-LOS configurations. These are used based on mission requirements, and are not normal wireless network extension configurations. (For more information on the AN/PSC-5 refer to TM 11-5820-1130-12&P.)

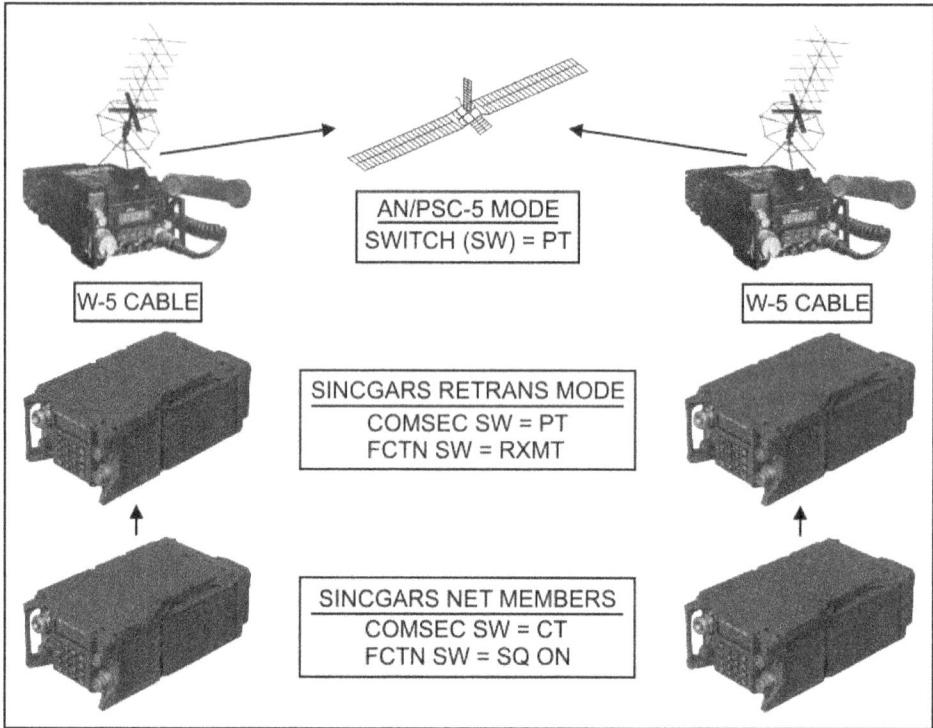

AN/PSC-5 MODE SWITCH (SW) = PT

W-5 CABLE   **W-5 CABLE**

SINCGARS RETRANS MODE
COMSEC SW = PT
FCTN SW = RXMT

SINCGARS NET MEMBERS
COMSEC SW = CT
FCTN SW = SQ ON

Figure 6-5. SINCGARS range-extension with Spitfire

COMMUNICATIONS PLANNING

6-29. The Network Management System (NMS) provides the joint staff, GCC planning facilities, and the subordinate units with a tool that consolidates and provides information to maintain a database. This database necessary for the controllers to implement the DAMA process and receive allocations of satellite resources.

6-30. The database will include information about the terminal, the user, and the services requested. This information will include, but is not be limited to—

- Type of terminal.
- COMSEC being employed (not including key type).
- I/O device attached.
- Data rates of the I/O device.
- Terminal address.
- Network addresses terminal.
- Guard lists.

Note. A guard frequency is a RF that is normally used for emergency transmissions and is continuously monitored for example, UHF band 243.0 MHz and VHF band 121.5 MHz.

6-31. Although the NMS will not replace the need to document requirements in the integrated communications database, it will eventually replace the need to generate a satellite access request. Information passes electronically from planners within the unit, to their higher headquarters, via the joint (UHF) military SATCOM network integrated control system as the mission dictates. When fielded, the Army's NMS will be a part of the integrated system control.

KEY DISTRIBUTION

6-32. Key distribution is critical in achieving secure satellite transmissions. The brigade COMSEC office of record is responsible for the brigade COMSEC account. It also provides logistical support for the control and distribution of internal brigade and subordinate battalion COMSEC material. Commanders must ensure these procedures are established in a unit SOP. (TB 380-41 provides information on the procedures for safeguarding, accounting, and the supply control of COMSEC material such as COMSEC material distribution.)

Joint Communications Security Key Distribution

6-33. A joint contingency force (JCF), corps, and division key management plan (KMP) provides guidance on the COMSEC key distribution; however, it does not change current unit procedures. The COMSEC custodian is responsible for KMP coordination and the frequency manager is responsible for the satellite access request. The COMSEC custodian and frequency manager need to ensure prior coordination is made between the two so that all requests for COMSEC have been identified for all units.

Transmission Security (Orderwire) Key Distribution

6-34. The DAMA KMP will provide guidance on obtaining orderwire keys using the EKMS with the DAMA control system. It will also provide instructions for the receipt of OTAR by the users. The Spitfire provides an OTAR capability for orderwire keys. Spitfire operators should have the current and next orderwire keys for each footprint in which they will be operating.

> *Note.* Only the requesting unit's COMSEC custodian with a valid COMSEC account can order these keys. (Refer to TB 380-41.)

6-35. The DAMA semi-automatic controller (and possibly the NCS) places the orderwire keys in positions 0–7; the Spitfire uses positions 1–8. Careful coordination must be performed before the execution of any DAMA operations. Additionally, the location of the key must be coordinated within each footprint to ensure compatibility with the controller in all AOs.

AN/PSC-5I UHF TACTICAL GROUND TERMINAL (SHADOWFIRE)

6-36. The AN/PSC-5I is a field upgrade of the AN/PSC-5 Spitfire terminal. The upgrade was designed to provide all the capabilities of the AN/PSC-5I plus additional capabilities for HAVEQUICK I and II and SINCGARS anti-jam; the ability to receive and transmit OTARs; extended 30–420 MHz frequency range, MIL-STD-188 to 181B HDR in LOS and SATCOM communications; and MIL-STD 188-184 embedded advanced data controller.

6-37. Additional features include embedded tactical Internet range extension and MELP voice coding, 142 preset channels, advanced key loading, DS-101 fill capability and embedded tactical IPs and COMSEC.

AN/PSC-5D MULTIBAND MULTIMISSION RADIO

6-38. The AN/PSC-5D offers a higher frequency range than the Spitfire and Shadowfire. A LOS, 5 kHz, 25 kHz DAMA, and 25 kHz SATCOM comparison of the AN/PSC-5 family of radios and the AN/PRC-117F is outlined in Table 6-1, 6-2 and 6-3. For more information on UHF SC TACSAT/DAMA refer to FM 6-02.90.

Table 6-1. AN/PSC-5/C/D, AN/PRC-117F and AN/ARC-231 LOS interoperability

Radio Item	AN/PSC-5 Spitfire	AN/PSC-5I Shadowfire	AN/PSC-5D and AN/ARC-231	AN/PRC-117F
Frequency Range MHz	30–400 MHz	30–420 MHz	30–512 MHz	30–512 MHz
Voice 12 kbps	FASCINATOR	FASCINATOR	FASCINATOR	FASCINATOR
Voice 16 kbps	VINSON	VINSON	VINSON	VINSON
Data 16 kbps	VINSON, 3 or 4 KG-84	VINSON, 3 or 4 KG-84	VINSON, 3 or 4 KG-84	VINSON, 3 or 4 KG-84
Data (over 16 kbps)	No	1–4 KG-84 (3 KG-84)—up to 48 kbps	1–4 KG-84 (3 KG-84)—up to 48 kbps	No
CTCSS	No	Yes	Yes	No
SINCGARS FH	No	Yes	Yes	Yes
Guard frequency	No	Yes	Yes	No
Channel Spacing	5 and 25 kHz	5 and 25 kHz	5, 6.25, 8.33, 12.5, 25 kHz	10 Hz, 5, 8.33, 12.5 and 25 kHz

Note. CTCSS—continuous tone coded squelch system

Table 6-2. AN/PSC-5/C/D, AN/ARC-231 and AN/PRC-117F
5 kHz and 25 kHz DAMA interoperability

Terminal / Mode	AN/PSC-5	AN/PSC-5C	AN/PSC-5D and AN/ARC-231	AN/PRC-117
5 kHz voice 2400 bps	ANDVT	MELP (AUTO)	MELP (AUTO)	MELP (AUTO)
5 kHz Data 2400 bps	ANDVT, 3 or 4 KG-84	ANDVT, 3 or 4 KG-84	ANDVT, 3 or 4 KG-84	ANDVT, 3 or 4 KG-84
5 kHz DASA (Data)	ANDVT, 3 or 4 KG-84 up to 2400 bps	1–4 KG-84 (3 KG-84)—up to 9600 bps	1–4 KG-84 (3 KG-84)—up to 9600 bps	1–4 KG-84 (3 KG-84)—up to 8000 bps *must use 181B for interoperability (HPW between 117F only)
25 kHz Voice 2400 bps	ANDVT	MELP (AUTO)	MELP (AUTO)	MELP (AUTO)
25 kHz Data 2400 bps	ANDVT, 3 or 4 KG-84	ANDVT, 3 or 4 KG-84	ANDVT, 3 or 4 KG-84	ANDVT, 3 or 4 KG-84
Data 4800 bps (limited access)	3 or 4 KG-84	3 or 4 KG-84	3 or 4 KG-84	3 or 4 KG-84
25 kHz DASA Data	Vinson, 3 KG-84/4 KG—84 only up to 16 kbps	1–4 KG-84 (3 KG-84)—up to 48 kbps	1–4 KG-84 (3 KG-84)—up to 56 kbps	1–4 KG-84 (3 KG-84)—up to 56 kbps *must use 181B for interoperability
Data transfer	Yes	Yes	Yes	No

Note. DASA—demand assigned single access   HPW—high performance waveform

Table 6-3. AN/PSC-5/C/D AN/ARC-231 and AN/PRC-117F
25 kHz SATCOM interoperability

Terminal / Mode	AN/PSC-5	AN/PSC-5C	AN/PSC-5D and AN/ARC-231	AN/PRC-117
Voice 16 kbps	VINSON	VINSON	VINSON	VINSON
Data 16 kbps	VINSON, 3 or 4 KG-84	VINSON, 3 or 4 KG-84	VINSON, 3 or 4 KG-84	VINSON, 3 or 4 KG-84
Data (over 16 kbps)	NO	1–4 KG-84 (3 KG-84)—up to 48 kbps	1–4 KG-84 (3 KG-84)—up to 56 kbps	1–4 KG-84 (3 KG-84)—up to 56 kbps *Must use 181B for interoperability (HPW between 117F only)

AN/PRC-117F MANPACK RADIO

6-39. The AN/PRC-117F is an advanced multiband/multimission manpack radio that provides reliable tactical communications performance in a small, lightweight package that can maximize user mobility. The AN/PRC-117F is a multiprocessor based, fully digital, software controlled, voice and data transceiver. The

AN/PRC-117F is capable of providing; LOS, SATCOM, ECCM, FH operations (SINCGARS and HAVEQUICK), and is compatible with all tactical VHF/UHF radios. (The AN/VRC-103 is the vehicular version of the AN/PRC-117F.) Refer to Figure 6-6 for an example of the AN/PRC-117F.

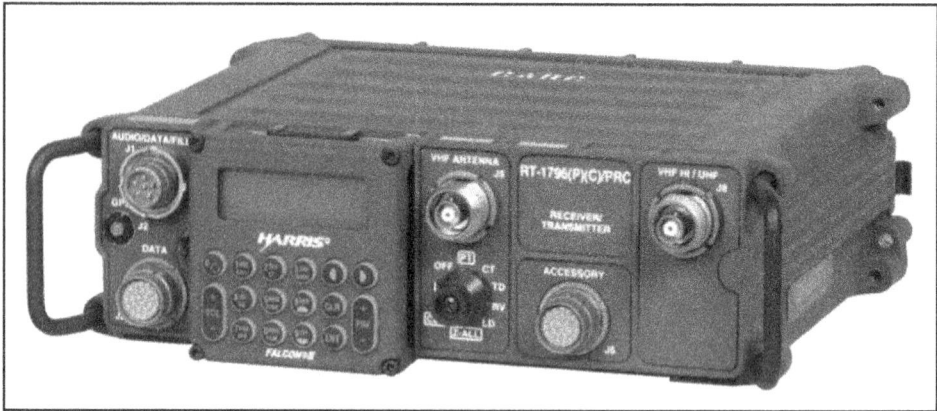

Figure 6-6. AN/PRC-117F

6-40. The AN/PRC-117F is a COTS radio and is covered under a commercial warranty. The radio requires regular updates to the firmware. Signal planners should pay special attention to ensuring that radios have the latest version, which is available from the Harris Premier Web site (https://www.premier.harris.com/), because having multiple versions of the firmware within a unit can cause problems with interoperability.

AN/PRC-117F CHARACTERISTICS AND CAPABILITIES

6-41. The AN/PRC-117F is designed to act as the transmission means for a range of command, control and communications input devices (both digital data and analog). These include standard audio (voice) communications via a handset; line-level audio-data devices such as the handheld data terminals found in SOF, military intelligence, field artillery and other units; analog teletype modems; C2 digital DTE as found in the ABCS; PCs; e-mail systems, video systems, fax and more. The AN/PRC-117F can operate across both the VHF and UHF military tactical frequency bands using either LOS modes or satellite propagation media for BLOS communications.

6-42. According to the article "*AN/PRC-117 Special Operations Forces Radio Has Applications for Digital Divisions and Beyond*", due to the microprocessor design, digital signal processing and software control, the AN/PRC-117F is actually the equivalent of many current radios in one manpack or vehicle mounted box. This greatly reduces the space, weight, power and support requirements for both individual fighting platforms and tactical-operations centers. This also greatly reduces co-site interference problems and, if used properly, can reduce the number of tactical radio nets required to support a digitally equipped fighting force. The AN/PRC-117F has the following characteristics and capabilities—

- **Frequency range of 30–512 MHz**. This frequency range covers not only the "standard" Army tactical (30–88 MHz) band but also covers the frequency bands and modulation modes commonly used by the USAF, USN and Coast Guard for operations, air traffic control (ATC), tactical data links and maritime uses. This makes the radio ideal for use as a "liaison radio" or "gateway" between service components using different waveforms for joint ground sea and air operations. Also, AN/PRC-117Fs frequency range and waveform modes are compatible with civil and public service frequency bands commonly used by non-DOD local, state, federal and foreign agencies.

- **Modulation**. As delivered, the radio is programmed at the factory for compatibility with current "standard" modulation characteristics segmented in the traditional RF bands—

    ▪ **VHF low band**. 30.00000–89.99999 MHz, FSK. This makes the radio interoperable with SINCGARS, AN/PRC-68, AN/PRC-126 and other tactical radios of both foreign and domestic manufacture.

    ▪ **VHF high band**. 90.00000–224.99999 MHz FM, AM, FSK, amplitude shift-keying. In this frequency band, the radio can be used for air-to-air, air-to-ground and ground-to-ground voice and data communications using waveforms found in this band. The AN/PRC-117F is compatible with a variety of existing military aircraft and air-traffic-control radio communications, as well as military air-to-ground data-link communications, the commercial USMC band, USN/Coast Guard communications and civil police, fire and emergency-management standard radios. Due to its capability, joint and civil-military liaison for both voice and data can be accomplished in one radio by units that have AN/PRC-117F. This is particularly important to the Army National Guard because of their large role in civil support operations.

    ▪ **UHF band**. 225.00000–511.99999 MHz. AM, FSK, amplitude shift keying. In this frequency band, AN/PRC-117F can be used to perform air-to-air, air-to-ground, ground-to-ground, fixed or mobile radio communications missions for both voice and data modes. The AN/PRC-117F is also compatible with ECCM-capable equipment such as AN/ARC-164 and AN/ARC-182 that can be widely found in existing tri-service ground, airborne and special-mission systems.

    ▪ **UHF SATCOM**. 243.00000–270.00000 MHz and 292.00000–318.00000 MHz. In this frequency range, AN/PRC-117F is fully compatible with SC and DAMA TACSAT systems. The AN/PRC-117F also has full orderwire capability and can send and receive data at a rate of 64 kbps in a 25 kHz channel or 12 kbps in a 5 kHz channel. Also, automatic requests for wireless network extension of bad data packets and COMSEC are embedded in the radio hardware and software. This key SATCOM capability gives the radio a feature no other standard CNR has: the ability to communicate BLOS without wireless network extension stations from the same radio package that's used for LOS communications.

6-43. The AN/PRC-117F operates in the following LOS fixed frequency CT operating capabilities and limitations—

- **VINSON**—16 kbps data rate, 25 kHz COMSEC (KY-57/58) mode for secure voice and data.
- **KG-84 compatible**—(data only) supports voice only using a 12 kbps data rate in FM and trellis code modulation from 30.00000–511.99999 MHz and AM mode from 90.00000–511.99999 MHz. Also available in all modes of UHF SATCOM.
- **TEKs**—electronically loaded 128 bit transmission encryption keys used to secure voice and data communications.
- **COMSEC fill**—TEKs, TSKs, and KEKs can be filled from the following devices—

    ▪ AN/CYZ-10, DTD (ANCD).
    ▪ AN/PYQ-10, SKL.
    ▪ KYK-13, electronic transfer device.
    ▪ KYX-15, net control device.
    ▪ MX-18290, ECCM fill device.
    ▪ KOI-18, general purpose tape reader.

6-44. The AN/PRC-117F can operate in HAVEQUICK I/II, utilizing FH from 225–400 MHz, providing compatibility with current airborne FH. It can also operate in SINCGARS FH mode from 30.0000–87.975 MHz. and supports SINCGARS SIP/ESIP features by being placed in either a net master or a net member mode.

6-45. The AN/PRC-117F can scan up to 10 LOS fixed frequency or dedicated SATCOM radio voice operation nets. It does not scan HAVEQUICK, SINCGARS, or UHF DAMA nets and digital squelch cannot be used. Scanning combinations of CT and PT nets is allowed by the PT override feature of the VINSON and FASCINATOR CT mode.

AN/PRC-117F DATA CAPABILITIES

6-46. The AN/PRC-117F can be used as a digital data-transmission device. The recommended standard-232 and 422, and MIL-188-114 I/O ports are provided integral to the radio, along with synchronous and asynchronous data interfaces. This makes it very easy to interface DTE, computer workstations and networking components such as, CP routers, to the radio for data transmission applications. The AN/PRC-117F can send data transmission rates of 56 kbps through SATCOM and 64 kbps ground-to-ground (LOS).

6-47. With these data rates, the AN/PRC-117F would make data transmission among brigade and battalion TOCs and lower echelons fast enough to support lengthy database-to-database transfers. Transmission of databases, plans, orders and reports that are now difficult and time consuming to do over tactical radios would be much faster. This would not only improve operations but would also reduce system vulnerability to enemy intercept and detection. Also, these rates will support user desired C2 tools such as video teleconferencing, imagery transmission, en route mission planning and collaborative planning that aren't practical using current lower-data-rate equipment. (Refer to Appendix G for more information on data communications.)

ARMY CONVENTIONAL FORCES

6-48. The primary mission of the Warfighter Network is to augment the current and projected C2 system. This system must always be operational to support requirements during peace, crisis, and war. The addition of the Warfighter Network ensures a C2 communications system across the operational continuum. CNR provides the commander the ability to immediately access CPs while operating on the move, eavesdrop on subordinate units' communications, and affect operations during critical moments of the fight.

6-49. The significant advantages of SINCGARS and SC TACSAT systems are to make them the recommended CNR communications means for the Warfighter Network. SINCGARS will continue to be widely available on the battlefield, easy to use, and interoperable with aircraft radio versions. It provides improved immunity from the EW threat. SC TACSAT terminals provide users with critical C2 connectivity over extended ranges. The Warfighter Network requires assured space segment access 365 days a year to support operational training. It ensures a smooth transition from peacetime operation to war. This assured access requirement is critical to force deployment operations, and equals USN and USAF requirements.

OPERATIONS AND INTELLIGENCE NETWORKS

6-50. The corps and division system improves intelligence planning, streamlines the handling of information, and expedites production of intelligence. Its purpose is to speed the flow of information up and down the chain of command using dedicated and secure communications nets. It also ensures the integration of information from all sources into a clear, accurate, and complete picture for the commander.

6-51. The corps combines intelligence and combat information from corps subordinate units and national/strategic, theater, combatant command, and multinational intelligence efforts. Fully integrated, all-source intelligence is produced at corps and is the basis of the commander's intelligence preparation of the battlefield.

6-52. SC TACSAT is a valuable communications asset for sustainment units in support of dispersed forces across full spectrum operations and communications zone. The requirement for SC TACSAT assets exists for sustainment units, from early entry, through normal daily operations in a mature theater. Units at all levels rely heavily on a fully planned and reliable communications architecture to provide SA, multimedia services, imagery, and asset visibility.

6-53. Additionally, the ability to access timely materiel and movement related information allows the logistician to focus on the discipline of distribution from the strategic, operational, and tactical echelons of logistical support, to sustain operations. The theater sustainment command, battlefield distribution, and velocity management concepts support the requirement for SC TACSAT capabilities. The SC TACSAT communications assets can provide continuous information feeds to Army and joint total assets visibility, which will achieve the leap-ahead capability that is necessary to support the Army's transformation to modularity.

SINGLE-CHANNEL TACTICAL SATELLITE FIRE SUPPORT NETWORKS

6-54. Doctrinally, most of the SC TACSAT nets used in the distribution plan for the Spitfire are voice nets. The need for a digital link between the Advanced Field Artillery Tactical Data System (AFATDS), Initial Fire Support Automation System, Forward Observer System, and non-fire support C2 systems may require these nets to be used for digital traffic. Voice/Data contention does not satisfy the requirements of fire support. The commander must decide which net will provide voice service, and which will carry data. These nets can be used for either voice or data, but not both.

CORPS FIRE SUPPORT NET

6-55. The purpose of the corps fire support net is for clearing fires, which refers to the coordination necessary when firing into an adjacent AO controlled by someone else. The coordination ensures the area is under enemy control and there are no friendly forces in the area. The primary users of the net include any of the following—

- Corps fires cell.
- Fires brigades.
- Armored cavalry regiment fires cell.
- Attack regiment fires cell.

FIRES BRIGADE COMMAND OPERATIONS NET

6-56. The fires brigade command operations net will contain the operations elements from the fires brigade, field artillery brigade, fires battalion, and Multiple Launch Rocket System (MLRS) battalions. The primary purpose of this net is to provide a long range C2 link to subordinate field artillery elements. This net is primarily a voice net, but can transmit digital traffic between AFATDS or other automated devices.

MLRS BATTALION COMMAND OPERATIONS NET

6-57. The MLRS battalion and battery fire direction centers will use the MLRS battalion command operations net to facilitate BLOS communications between the MLRS battalion and its subordinate batteries. While primarily a voice net, the MLRS battalion command's operations net may be designated as a digital net, used to transmit AFATDS traffic.

DIVISION FIRE SUPPORT NET

6-58. The principle members of the division fire support net include the division fires cell, fires brigade, the brigade fires cell, fires battalion and the MLRS battalion. This net is used for fire support coordination and as an alternate for fire direction with elements throughout the division. The division TOC is typically the NCS. This net will normally operate as a voice net.

6-59. The separate brigade has unique long haul communications requirements, which LOS operations cannot satisfy when dispersed over extended distances. These units deploy UHF SC TACSAT terminals with their headquarters to provide C2 connectivity with higher headquarters. The primary communications mode is secure voice.

AIRBORNE AND AIR ASSAULT UNITS

6-60. The airborne and air assault units have a need for en route communications to maintain a connection with the sustaining base, other aircraft, and with the units that may already be in place. This is accomplished by using a secure en route communications package (SECOMP), which uses the Spitfire or a

VHF/UHF DAMA-capable SC TACSAT. The DAMA-capable SC TACSAT will provide communications in both the LOS and SATCOM modes. The SECOMP supports the commander and his principal staff while in route to the AO. It supports ground operations independently of the aircraft at staging areas and during joint task force (JTF) initial ground operations.

Secure En Route Communications Package

6-61. The SECOMP provides Army JCF with the necessary real-time communications to receive orders from higher headquarters. It allows the JCF to plan, coordinate, and rehearse mission operations, and to receive and disseminate near real-time, up-to-date intelligence information. The system provides Soldiers en route to an AO with the ability to communicate both vertically and horizontally from higher to lower, inter-aircraft, intra-aircraft, inter-service, and air-to-ground with both multinational and joint forces.

6-62. The SECOMP is connected via the coaxial cable into the aircraft satellite antenna system, or into an unused aircraft UHF antenna for LOS operations. The RTO, under the aircrew's supervision, connects/disconnects the radio from the aircraft antenna cable system. The RTO will remove the radio when exiting the aircraft. (Refer to Appendix A for more information on FM networks.)

SINGLE-CHANNEL TACTICAL SATELLITE COMMUNICATIONS PLANNING

6-63. Tactical communications networks change constantly. Unless control of the network is exercised, it will result in communications delay and a poor grade of service. The best method of providing this control without hampering operation is through centralized planning. Execution of these plans should be decentralized. This concept is applied to the space systems portion and to the ground stations. The US military satellite systems consist of terminals (ground segment), satellites (space segment), and tracking, telemetry, and control terminals (control segment).

6-64. The planning and system control process helps communications systems managers react appropriately to the mission of the force supported, the needs of the commander, and the current tactical situation. The type, size, and complexity of the system being operated establish the method of control.

6-65. Communications control is a process in which the matching of resources with requirements takes place. This process occurs at all levels of the control and management structure. In each case, the availability of resources is considered.

Chapter 7

Airborne Radios

Airborne radios provide communications for ground-to-air operations as well as air-to-air and air-to-sea missions. This chapter addresses airborne SINCGARS, the AN/ARC-210, AN/ARC-220, AN/ARC-231, AN/ARC 164, AN/ARC-184, AN/VRC-100, AN/VRC-83, and the AN/ARC-186.

AIRBORNE SINGLE-CHANNEL GROUND AND AIRBORNE RADIO SYSTEMS

7-1. The following paragraphs address the airborne SINCGARS. (For more information on aviation brigades and communications refer to FM 3-04.111.)

AN/ARC 201

7-2. Ground and airborne versions are interoperable even though they are physically different from each other. The major change in the airborne mode is the faceplate that is attached to the different configurations plus the add-on modules change each version's capabilities. Airborne versions RT-1476/1477A/B/C require the TSEC/KY-58 security equipment for CT operation.

7-3. The RT-1476/ARC-201 (refer to Figure 7-1) is the base radio in all three versions and they all operate in both the SC and FH modes. The RT-1476/ARC-201 is controlled from the front panel. It is designed to be mounted in the cockpit of an aircraft. (Refer to TM 11-5821-357-12&P for more information on the AN/ARC-201.)

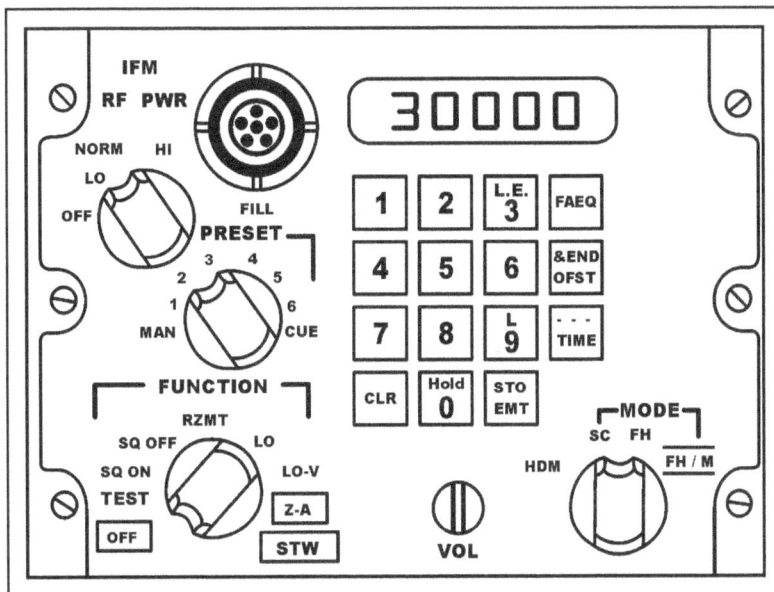

Figure 7-1. Airborne radio RT-1476/ARC-201

RT-1477/ARC-201

7-4.   The RT-1477/ARC-201 provides a remote capability for installations where the radio must be located away from the pilot's cockpit. It has a separate radio and a radio set control (also known as a RCU), C-11466, so the pilot can remotely control the radio from his position in the aircraft. All controls are on the RCU, located in the aircraft cockpit. The RT is located in a remote equipment compartment on the aircraft. Control and status signals are sent back and forth between RT and RCU via dedicated cables. The RT-1476/1477 has wireless network extension capabilities.

RT-1478/ARC-201

7-5.   The RT-1478/ARC-201 is a remote controlled RT. The aircraft system control display unit (CDU) controls the RT. The RT is located in the remote equipment compartment of the aircraft. The optional DRA, CV-3885/ARC-201, processes 1,200 and 2,400 Hz FSK data through the radio set for data transmission and interfaces between the RT and the TSEC/KY-58 COMSEC equipment. Operation of the DRA is automatic; there is no operator interface.

SINCGARS AIRBORNE SYSTEM IMPROVEMENT PROGRAM

7-6.   The SINCGARS airborne system improvement program (AIRSIP) contains throughput and robust enhancements. It includes a wireless network extension capability in the packet mode, improved error correction, more flexible remote control, and GPS compatibility. Additionally, the AIRSIP combines three line replaceable units (LRUs) (RT-1478, DRA, and external COMSEC/KY-58) into one unit, and reduces the overall weight of the radio system.

7-7.   The SINCGARS AIRSIP RT-1478D/ARC-201D is a VHF FM radio set that provides users with the ability to transmit and receive voice and data communications in the 30–88 MHz band. The integration of COMSEC and the DRA combines three LRUs into one enclosed system. The radio can operate in secure or PT mode. When operating in the FH mode, the radio provides an EP capability. The RT-1478D/ARC-201D provides voice interoperability with legacy radios in the SC mode and is fully interoperable with the SINCGARS family of ground and airborne radios. Figure 7-2 is an example of the SINCGARS AIRSIP, RT-1478D.

7-8.   The RT-1478D/ARC-201D key features include—

- Automatic wireless network extension.
- Built-in amplitude homing.
- Integrated DRA functions to include:
    - TACFIRE and SINCGARS data modes: 600, 1,200, 2,400, 4,800, and 16,000 bps.
    - Enhanced packet data modes: 1200N, 2400N, 4800N, 9600N; recommended standard-232 packet; and recommended standard-423 EDM is 16,000 bps only.
    - 1553B bus: provides both radio control and data I/O.
- BIT function.
- AM-7189/ARC compatible.
- Six FH presets (including TRANSEC keys).
- Six SC presets, plus manual and cue channels.

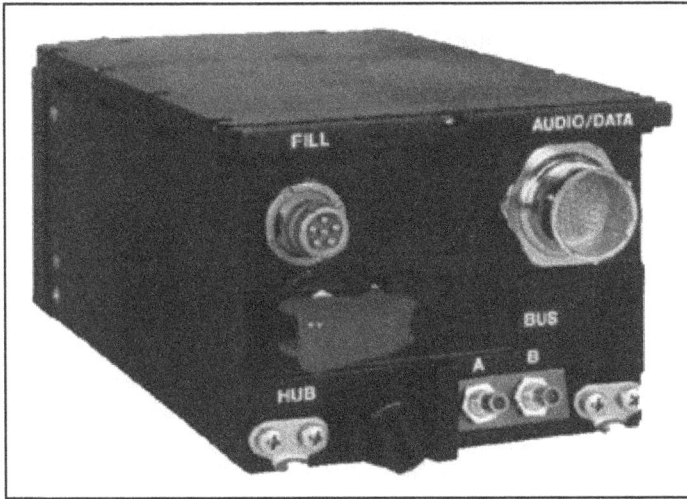

Figure 7-2. RT-1478D SINCGARS AIRSIP

AN/ARC-210 RADIO SYSTEM

7-9. The AN/ARC-210 is offered in several models, which when coupled with ancillary equipment, provides the aviation community with exceptional long range capability. The RT-1556B provides LOS VHF/UHF capability and HAVEQUICK I/II, and SINCGARS ECCM waveforms. The RT-1794I (refer to Figure 7-3), RT-1824I, RT-1851I, and RT-1851AI are network capable and include embedded COMSEC, 5 kHz and 25 kHz and DAMA SATCOM, and are certified to MIL-STD-188-181B/-182A/-183.

Figure 7-3. RT-1794 I

7-10. The AN/ARC-210 provides air-to-air and air-to-ground, two-way voice communication in both the UHF ranges and VHF. Data and voice communications are provided via the embedded SATCOM functions that operate in the UHF radio band.

7-11. The AN/ARC-210 provides the following key features—

- 30–400 MHz frequency range provides VHF and UHF in all radios; 121.5 and 243.0 MHz guard channels, and 4 channel scan.
- 30–512 MHz frequency range providing VHF and UHF in the RT-1851AI; 121.5 and 243.0 MHz guard channels, 4 channel scan.
- Synthesizer speed and rapid radio response time handles any developed ECCM algorithm or LINK requirement.
- Data rates up to 80,000 bps (SATCOM) and 100,000 bps LOS with bandwidth efficient advanced modulation technology.
- Compatible with Link 11, Link 4A and improved data modem.
- MIL-STD-1553B or remote control and BIT to module level.
- Channel spacing of—
    - 25 kHz (30–512 MHz).
    - 8.33 kHz (118–137 MHz).
    - 12.5 kHz (400–512 MHz).
- Tuning capability: 5 kHz with remote control, 2.5 kHz via 1553 bus.
- Optional PAs, mounts, and low noise amplifier/diplexer.

AN/ARC-220 RADIO SYSTEM

7-12. The AN/ARC-220 radio system is a microprocessor-based communications system intended for airborne applications and also has a ground version (AN/VRC-100). The AN/ARC-220 radio system uses advanced digital signal processor technology.

7-13. It consists of three replaceable units; a RT (RT-1749/URC or RT-1749A/URC), PA coupler (AM-7531/URC), and CDU (C-12436/URC). The AN/ARC-220 has embedded ALE, serial tone data modem, and anti-jam (ECCM) functions. The RT provides the electrical interface with other AN/ARC-220 LRUs and associated aircraft systems such as interphone, GPS, and secure voice systems. It also offers the ability for up to 25 free text data messages to be pre-programmed via data fill or created/edited in real time and the ability to receive data messages to be stored for later viewing.

7-14. The AN/ARC-220 radio system is capable of wireless network extension if desired and built-in integration with external GPS units allow position data reports to be sent with the push of a button. For more information on the AN/ARC-220 radio system refer to TM 11-5821-357-12&P.

7-15. The AN/ARC-220 radio system (refer to Figure 7-4) provides the following capabilities—

- Frequency range from 2.000–29.9999 MHz in 100 Hz steps.
- 20 user programmable simplex or half duplex channels.
- 20 programmable simplex or half duplex channels.
- 12 programmable ECCM hop sets.
- Certified for ALE in accordance with MIL-STD-188-141B and MIL-STD-188-141B.
- An integrated data modem which enables communication in noisy environments where voice communications are often not possible.
- Built-in integration with external GPS units allows position data reports to be sent with the push of a button.
- Embedded ALE, ECCM, and data modem (Joint Interoperability Test Command certified).
- Ability to rapidly and efficiently tune a variety of antennas.

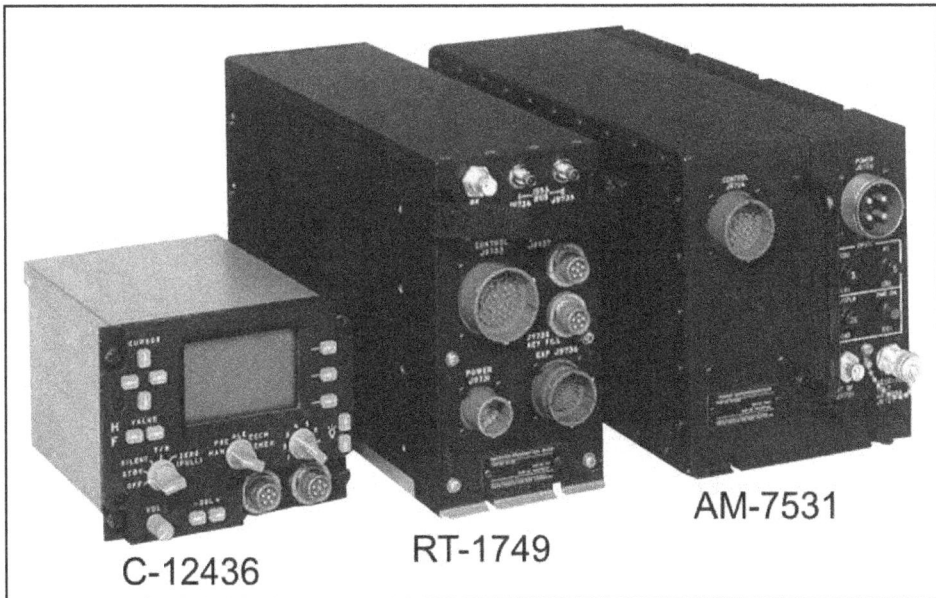

Figure 7-4. AN/ARC-220 radio system

AN/VRC-100(V) HIGH FREQUENCY GROUND/VEHICULAR COMMUNICATIONS SYSTEM

7-16. The AN/VRC-100(V) ground radio uses the RT, PA/coupler, and CDU LRUs of the AN/ARC-220 system without modification, within an aluminum-structured, bracketed case. It has a portable, metal case, with a removable top, that provides easy access for removal of LRUs. All controls, and the radio I/O, are located on the front panel. The AN/VRC-100 is intended for use in TOCs, ATC, and vehicular applications such as the high mobility multipurpose wheeled vehicle. Its key features are—

- Full digital signal processing with embedded ALE, EP, and data modem.
- Spare card slot in the RT provides for future growth.
- Operates on 28 VDC (and is compatible with 24 VDC vehicular power) or from 115 or 220 volts alternating current (VAC) 50/60 Hz power source.
- PC or laptop connectivity.
- E-mail messaging using local recommended standard-232 interface.
- Ability to effectively tune a variety of antennas.

7-17. Table 7-1 lists the three basic configurations of the AN/VRC-100 and Figure 7-5 is an example of an AN/VRC-100(V). Refer to TM 11-5820-1141-12&P for more information on the AN/VRC-100(V) 1/2/3.

Table 7-1. AN/VRC-100 configurations

Configuration	Description
AN/VRC-100(V) 1	Consists of three LRUs housed in a metal casing with a power supply and speaker.
AN/VRC-100(V) 2	Consists of the AN/VRC-100(V) 1 mounted in a wheeled vehicle.
AN/VRC-100(V) 3	Consists of the AN/VRC-100(V) 1 with the AS-3791/G broadband antenna and is used at theater level.

Figure 7-5. AN/VRC-100(V) high frequency radio

AN/ARC-231 RADIO SYSTEM

7-18. The AN/ARC-231 (refer to Figure 7-6) is an airborne VHF/UHF LOS and DAMA SATCOM radio system that is also a multiband/multimission, secure anti-jam voice, data and imagery radio set. The RT-1808 is the primary radio for the AN/ARC-231. One of the key features of the RT 1808 is that it capitalizes on the AN/PSC-5 Spitfire's expandable modular architecture and permits users to upgrade as new requirements drive new capabilities. The AN/ARC-231 is being used in the A2C2S to provide C2 mission capabilities to corps, division maneuver brigade, or attack helicopter commander's airborne TAC CP.

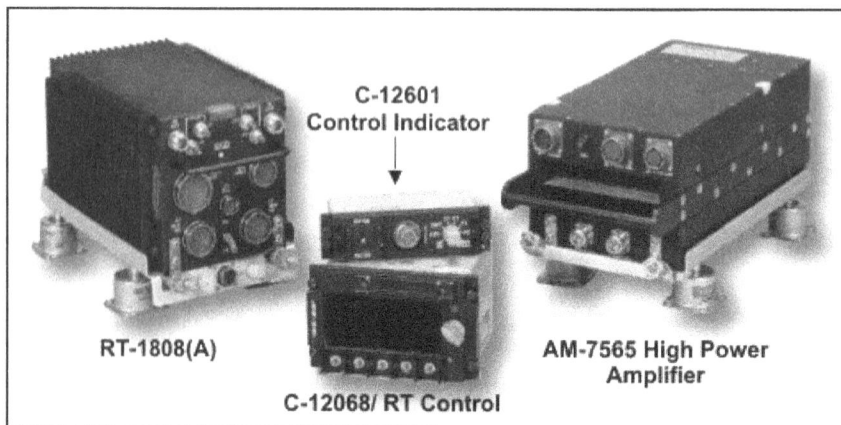

Figure 7-6. AN/ARC-231 radio system

7-19. The AN/ARC-231 has the following characteristics and capabilities—
- HAVEQUICK I/II and SINCGARS communications modes.
- DAMA and non-DAMA SATCOM communications modes.
- Frequency ranges of:
    - 30–87.975 MHz VHF FM SINCGARS.
    - 108–173.995 MHz VHF AM and VHF FM.
    - 225–399.995 MHz UHF AM HAVEQUICK II/ground air band, UHF SATCOM band.
    - 403–511.995 MHz UHF FM public service band.
- Embedded COMSEC and TRANSEC keys with transmit and receive OTARs.
- 148 preset channels.
- Independent red and black MIL-STD-1553 bus interfaces.
- Embedded MIL-STD-188-184 analog to digital converter and tactical IPs.
- SINCGARS SIP and optional ESIP and end of message.
- MIL-STD-188-181B high data rate in both LOS and SATCOM.
- 8.33 kHz ATC channelization coverage to 512 Hz.
- Minimal size and weight suitable for rotary and fixed wing applications.

AN/ARC-164(V) 12 ULTRA HIGH FREQUENCY RADIO

7-20. The AN/ARC-164(V) 12 radios are used for air-to-air, air-to-ground and ground-to-air communications. There are three major aircraft configurations of the AN/ARC-164 radio and one ground configuration of the AN/VRC-83(V). The AN/ARC-164(V) 12 RT configurations include a panel mount (RT-1518C), remote control (C-11721), remote mount (RT-1504) (refer to Figure 7-7), and data bus compatible (RT-1614). These radios provide anti-jam, secure communications links for JTF and Army aviation missions. The Army operational forces utilizing these radios are aviation units, air traffic services and Ranger units. It also provides the Army the ability to communicate with USAF, USN and NATO units in the UHF-AM mode which is the communications band for tactical air operations.

7-21. The AN/ARC-164(V) 12 has the following capabilities and characteristics—
- Operations in SC or FH mode.
- Frequency range of 225–399.975 MHz.
- Capacity of 7,000 channels.
- Embedded ECCM anti-jamming capabilities.
- Voice and data modulated signals with VINSON or VANDAL devices.

7-22. Refer to FM 6-02.771 for more information on HAVEQUICK radios and TM 11-5841-286-13 for information on the AN/ARC-164(V) 12.

Figure 7-7. RT-1504 for an AN/ARC-164(V) 12

AN/VRC-83(V) RADIO SET

7-23. The AN/VRC-83(V) is a two-band VHF AM and UHF AM radio set. The AN/VRC-83(V) is designed for tactical short range ground-to-ground and ground-to-air communication. The AN/VRC-83(V) is ground configuration of the AN/ARC-164, which is described in the next section. The AN/VRC-83(V) can operate in the jam-resistant, ECCM mode or in the NORMAL (non-ECCM) mode and can be used with COMSEC TSEC/KY-57 speech security equipment for secure voice communication.

7-24. The AN/VRC-83(V) is tunable in 25 kHz steps to either one of two frequency bands, VHF (116.000–149.975 MHz with 1360 channels) or UHF (225.000–399.875 MHz with 7000 channels). The AN/VRC-83(V) also has an RF PA to increase the RT transmit power of the set, an audio amplifier, a power supply to regulate the input voltage, a speaker and a handset. The handset is the audio input-output device for the radio set.

7-25. Primary components of the AN/VRC-83(V) consist of: one RT-1319B/URC, radio amplifier AM-7176, VRC-83 mount, cable assemblies, and a handset. Refer to Figure 7-8 for an example of the AN/VRC-83(V) and TM 11-5820-1149-14&P for more information on radio maintenance.

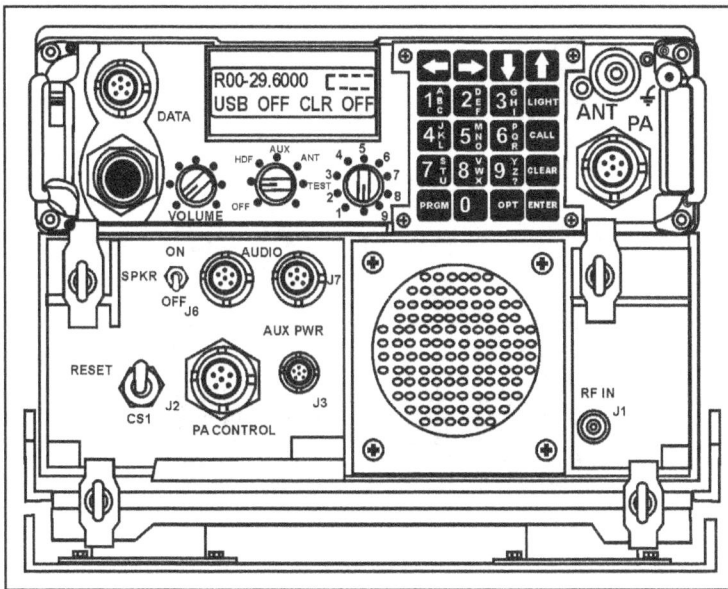

Figure 7-8. AN/VRC-83 radio set

AN/ARC-186(V) VHF AM/FM RADIO

7-26. The AN/ARC-186(V) provides AM, FM, FM homing and wireless network extension. It is primarily used as an administrative VHF AM/FM radio used to communicate with the ATC. The AN/ARC-186(V) is a LOS radio system with limited range at terrain-flight altitudes but greater range at administrative altitudes normally associated with ATC communications. It can back up the SINCGARS in the same 30–89.975 MHz frequency range but a big disadvantage is that it has no FH mode compatible with SINCGARS and it generally lacks KY-58 interface to provide secure FM communications.

7-27. Battalions typically operate a C2 network, O&I and A&L network all using SINCGARS. Battalions also operate an internal air operations network using HAVEQUICK II. The AN/ARC-186(V) is a secondary means of secure tactical communication to overcome SINCGARS and HAVEQUICK II LOS constraints.

7-28. Even though the AN/ARC-186(V) VHF AM radio is normally used for administrative purposes it may function as a platoon internal net. The battalion TOC may also have access to MSE and SATCOM for communicating with higher headquarters. (Refer to TM 11-5821-318-12 for more information on the AN/ARC-186(V).)

7-29. The AN/ARC-186(V) (refer to Figure 7-9) has the following capabilities—
- Secure communications when the radio is employed with the KY-58.
- Frequency ranges of:
    - AM transmit/receive: 116–151.975 MHz.
    - AM receive only: 108.000–115.975 MHz.
    - FM transmit/receive: 30.000–87.975 MHz.
- Channel spacing: 25 kHz.
- 20 preset channels with electronic memory.

Figure 7-9. AN/ARC-186 (V)

Chapter 8

Other Tactical Radio Systems

To be successful on the modern battlefield, commanders must be able to communicate in order to control and coordinate movement, send and receive instructions, request logistical or fire support, and gather and disseminate information. In addition to CNR systems, many other tactical radio systems are now available. The means of communications chosen will depend on the situation. This chapter addresses the AN/PRC-126, ICOM F43G, the LMR, the Land Warrior (LW) communications networking radio subsystem (CNRS), combat survivor evader locator radio, and the JTRS.

AN/PRC-126 RADIO SET

8-1. The AN/PRC-126 (refer to Figure 8-1) is susceptible to adversary jamming and friendly co-site interference. Alternate frequencies must be identified for use in case of jamming, and leaders must ensure that Soldiers are trained to recognize, overcome, and report jamming activities.

8-2. The AN/PRC-126 enables small unit leaders to adequately control the activities of subordinate elements in accomplishing the unit's mission. It is a short-range, handheld, or vehicular mounted tactical radio, used primarily at the squad/platoon level. Vehicular power requires connection to an OG-174, amplifier power supply. It's key features include—

- Lightweight, militarized transceiver providing two-way voice communications.
- Frequency range of 30–87.975 MHz.
- Frequency separation is 25 kHz.
- Nominal range for reliable communications over rolling, slightly wooded terrain is 500 meters (1,640.4 ft) with the short antenna, or 3,000 meters (9,842.5 ft) with the long antenna.
- Standard battery (lithium) operating time is 70 hours.
- Capable of operating with SINCGARS in the fixed frequency mode.
- Capable of providing secure voice operation when used with the TSEC/KYV-2A secure voice module.
- Digital communications for passing TACFIRE data are possible when connected to the OG-174. (Refer to TM 11-5820-1025-10 for more information on the AN/PRC-126 and FM 6-50 for additional information on transmitting TACFIRE data with the AN/PRC-126.)

8-3. In the light infantry platoon, the rifle squad has two AN/PRC-126 radios: one for the squad leader and the other for the A-team leader. Air assault and airborne infantry squads have only one AN/PRC-126 each. If tasked to conduct a patrol, the dismounted section of a Bradley infantry fighting vehicle mechanized infantry platoon, should task organize its radio equipment in the preparation phase to ensure teams will have communications.

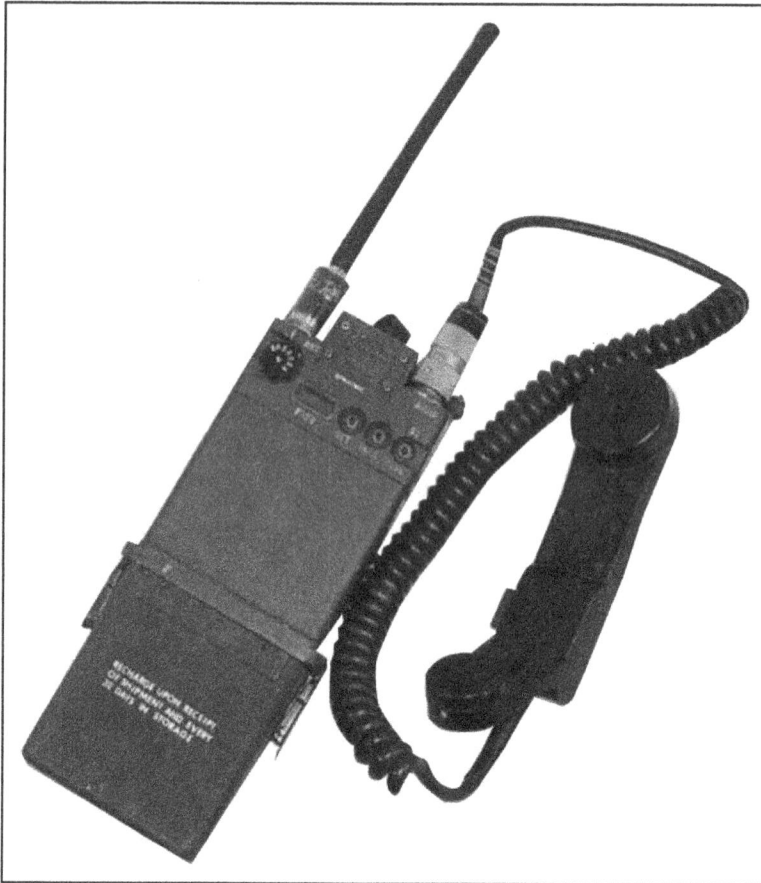

Figure 8-1. AN/PRC-126 radio set

ICOM F43G HANDHELD RADIO

8-4. The ICOM F43G handheld radio is a COTS system. It is a short range, handheld radio fielded with headset and an encryption module. It is employed at the lowest echelon of command, to control squads and teams. The ICOM F43G is used to provide the Soldier with a small light weight, rugged handheld radio with capability of secure UHF 2-way communication.

8-5. The ICOM F43G (Figure 8-2 is an example of the ICOM F43G) has the following characteristics and capabilities—

- UHF operation in the 380–430 MHz frequency range.
- 4 miles (6.4 km) plus transmission in the unencrypted mode.
- 256 memory channel capacity.
- 16 memory banks that allow for division and storage of a variety of flexible channel groupings.
- Built-in multi-format tone signaling and built-in voice scrambler.
- It uses a data encryption standard card which can be upgraded for secure communication.

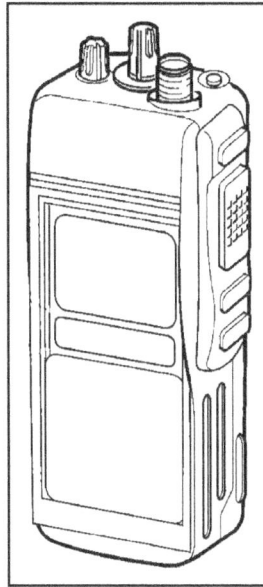

Figure 8-2. ICOM F43G handheld radio

LAND MOBILE RADIO

8-6. The LMR is typically the primary system used for daily installation communications. It is also commonly employed and used for administrative installation activities in public safety organizations and is compliant with the Association of Public Safety Communications Officials (normally referred to as APCO) Project 25 (P25) standards. P25 standards are based on the public safety communities needs as they define them. The LMR enhances communications interoperability with state and local agencies in a homeland defense or disaster situation.

8-7. LMR systems range from SC analog to digital trunked systems. The most basic LMR systems are SC analog systems. Each radio is set to a particular frequency that must be monitored by everyone utilizing the same channel. These systems have a dedicated channel for each group or agency using the system. In smaller agencies, if the system experiences heavy usage, users may not be able to place calls. The majority of these systems are VHF systems that offer very little flexibility in their operations. These systems fail to provide a common air interface and cannot accommodate users outside the system. These systems are inefficient users of spectrum, and many agencies have outgrown them. For United States and Possessions (US&P) LMR regulations see Chapter 8 of the National Telecommunications and Information Administration (NTIA) *Redbook*.

8-8. The majority of public safety organizations are currently using SC analog systems. Many of these organizations are in the process of switching, or have switched to, digital trunked systems. Trunked systems utilize a relatively small number of paths, or channels, for a large number of users. This is similar to commercial telephones. Rather than having a dedicated wire line for every user, the phone company has a computer (switch) that manages many calls over a relatively small number of telephone lines. This is based on the assumption that not every user will require a line at the same time.

8-9. Trunked systems are generally made up of a control console, repeaters, and radios. Instead of using switches and phone lines, these systems use consoles and channels or frequencies to complete calls. The process is the dynamic allocation of a channel that is totally transparent to the user. When the user of a trunked system activates the push-to-talk, the system automatically searches for an unused channel on which to complete the call.

8-10. Digital trunked systems offer better performance and provide a more flexible platform. This system accommodates a greater number of users and offers an open ended architecture. This allows for various modes of communications such as data, telephone-interconnect, and security functions. Additionally, there is faster system access, more user privacy, and the ability to expand by providing a common air interface. For CONUS LMR regulations refer to the NTIA *Redbook*, Chapter 10. The user/unit is responsible for obtaining a frequency assignment IAW NTIA, Manual of Regulations and Procedures for Federal Radio Frequency Management; AR 5-12; AR 25-1; and FM 6-02.70. Operation of Radio Frequency System without spectrum authorization/assignments is prohibited. (Refer to Figure 8-3 for an example of the LMR.)

8-11. The LMR has the following characteristics and capabilities—

- Frequency range of 380–470 MHz.
- Power of 1–4 Watts.
- Battery life of 10 hours.
- Secure (National Institute of Standards and Technology Type III) point-to-point voice communications.
- Range of 5 km (3.1 miles) max over smooth terrain.
- Programming of up to 512 channels.
- Easy radio reprogramming feature.
- Immersible to a depth of 1 meter (3.2 ft) for 30 minutes.
- Supports both narrowband (12.5 kHz) and wideband (25 kHz) channel spacing.
- Intra-Squad/Team Communications for non-critical C2, admin and logistics functions.

Figure 8-3. Land mobile radio

LAND WARRIOR

8-12. The LW, Figure 8-4, is a ground Soldier system, which integrates everything an infantry Soldier wears or carries on the battlefield. It is based on advances in communications, sensors, and materials. The LW integrates commercial technologies into a complete Soldier system. Its components include a: helmet subsystem, weapons subsystem, Soldier control unit, power subsystem, navigation subsystem, computer subsystem and CNRS.

Note. Information in this manual was current at time of publication and the LW had only been fielded to 4/9th Infantry Regiment.

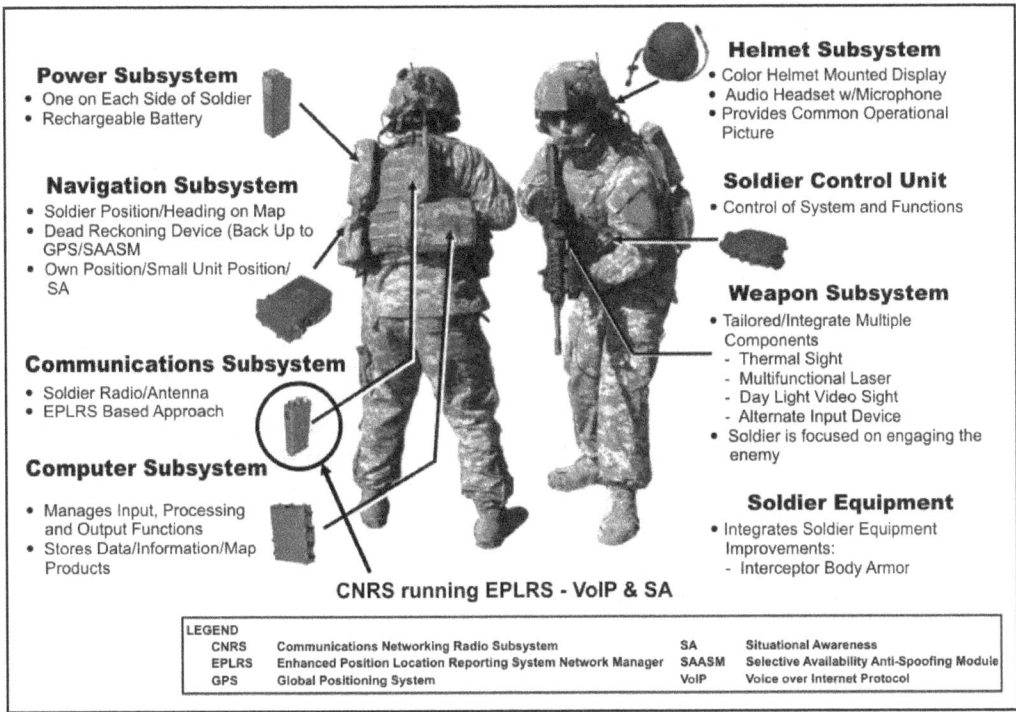

Power Subsystem
- One on Each Side of Soldier
- Rechargeable Battery

Navigation Subsystem
- Soldier Position/Heading on Map
- Dead Reckoning Device (Back Up to GPS/SAASM)
- Own Position/Small Unit Position/ SA

Communications Subsystem
- Soldier Radio/Antenna
- EPLRS Based Approach

Computer Subsystem
- Manages Input, Processing and Output Functions
- Stores Data/Information/Map Products

Helmet Subsystem
- Color Helmet Mounted Display
- Audio Headset w/Microphone
- Provides Common Operational Picture

Soldier Control Unit
- Control of System and Functions

Weapon Subsystem
- Tailored/Integrate Multiple Components
  - Thermal Sight
  - Multifunctional Laser
  - Day Light Video Sight
  - Alternate Input Device
- Soldier is focused on engaging the enemy

Soldier Equipment
- Integrates Soldier Equipment Improvements:
  - Interceptor Body Armor

CNRS running EPLRS - VoIP & SA

LEGEND
CNRS — Communications Networking Radio Subsystem
EPLRS — Enhanced Position Location Reporting System Network Manager
GPS — Global Positioning System
SA — Situational Awareness
SAASM — Selective Availability Anti-Spoofing Module
VoIP — Voice over Internet Protocol

Figure 8-4. Land Warrior

COMMUNICATIONS NETWORKING RADIO SUBSYSTEM

8-13. The CNRS offers the same functionality as the full size EPLRS radio (refer to Chapter 5 for more information on EPLRS radios), except that it uses a smaller, lighter, and more power efficient EPLRS, the RT 1922 C/G. It also supports multiple simultaneous channels of contention free voice (Voice over IP) and data. Power is supplied through the radio's interface connector pins, using an external (remote) battery pack or an external power supply.

8-14. The CNRS characteristics and capabilities include—
- Voice and data communications in a single streamlined unit.
- Interoperability with EPLRS and JTRS and integrates with the Army's tactical Internet FBCB2.
- LOS with automatic hopping within the tactical Internet for range extension.
- Handling of SECRET and "Secure But Unclassified" material.
- 2.9 megabits per second per network with a maximum of 486 kbps per user.
- Power out of 50 milliwatts to 5 watt (batter pack).
- Weight of approximately 12 ounces (without the power supply).

Land Warrior/CNRS Bandwidth Methodolgy

8-15. The EPLRS is an adaptive mobile radio network for users within tactical units. It employs embedded security features and automatic relaying which is used for range extension and is transparent to the user. EPLRS uses TDMA to allow a user to participate simultaneously in more than one network. The EPLRS TDMA architecture is divided into eight logical time slots (LTS). Each separate net (for example: battalion SA, battalion other data, company voice) is assigned its own LTS, or portion of a LTS, and frequency.

8-16. The EPLRS offers three types of communication services: duplex, CSMA, and MSG. The LW SA and C2 nets are CSMA allowing everyone on the net the capabilities of transmitting and receiving all traffic transmitted/received on the net.

8-17. The architecture allocates a one-half LTS CSMA short local needline for the battalion position report SA net. The short indicates that all messages must fit into one transmission unit (648 bits); local indicates the messages will be relayed one time; and one-half LTS provides a user data rate of 9.4 kbps.

8-18. The number of participants in the battalion position report SA net that will receive the position report SA updates from others on that net is dependent on distribution and actual ranges. When the battalion is distributed over a 30 km (18.6 miles) area, approximately 45 percent of the battalion will be within range of each other (including radio relay).

8-19. Unlike the FBCB2 architecture, LW does not use SA servers to forward information. The LW battalion data needlines are two hop nets. Position reports SA from key LW roles is forwarded to the lower tactical Internet SA net which is then relayed 4 hops and retransmitted down to LWs beyond the two hop limit of the LW data needlines as part of the lower tactical Internet brigade SA on the other LW needline. This ensures that most of the LW systems will consistently receive the updated position report SA for key LW roles. It is considered acceptable if a LW periodically misses an SA update because of the frequency of updates. A C2 message will be resent up to three times before the LW system stops trying to send the message.

8-20. For each unit on the SA network, an SA message is sent either at regular time intervals when position is unchanged (the unit is stationary), or when position changes by a specified distance. This is known as a time-motion update rate (distance). For example, the time-motion update rate for the company vehicles is 300 seconds or 100 meters (328 ft). The current LW default settings are two minute updates if stationary, every 15 meters (49.2 ft) of movement, or every 15 seconds if moving faster than 60 meters (196.8ft)/minute.

8-21. Knowing the number of EPLRS equipped company vehicles and LW platforms, and using assumptions about the time-motion update rates and the number of platforms moving, the number of messages that would be sent in one hour can be calculated. SA messages are 496 bits, which is smaller than the EPLRS transmission unit of 648 bits. Since a transmission unit represents the smallest amount of information that can be transmitted, each SA message sent will consume 648 bits.

8-22. From this information, the offered data load for one hour will be summed and the average bps determined. This average divided by the data rate of the EPLRS radio produced a utilization number for the network. Previous modeling and testing of the EPLRS network have developed a relationship between network utilization and message completion rate. This relationship will be used to estimate the message completion rate for each LW EPLRS net.

COMBAT SURVIVOR EVADER LOCATOR

8-23. The combat survivor evader locator (CSEL) handheld radio is utilized for locating and rescuing downed aircrew members. It is primarily used by personnel assigned as flight crews, SOF and other personnel with a high priority of becoming isolated. The CSEL is the primary search and rescue system used by SOF and aviation units. Its enhanced capabilities are not available by the older transceivers; AN/PRC-90 and AN/PRC-112.

8-24. The CSEL system is composed of three segments: over-the-horizon segment, ground segment, and the user segment. The three segments use GPS, national and international satellites and other national systems to provide geopositioning and radio communications for personnel recovery.

OVER-THE-HORIZON SEGMENT

8-25. The over-the-horizon segment operates over UHF SATCOM systems and Search and Rescue Satellite Assisted Tracking. The UHF SATCOM mode supports two ways messaging/geoposition between an AN/GRC-242 radio set base station and the AN/PRQ-7 radio set.

GROUND SEGMENT

8-26. The ground segment is composed of CSEL workstations and the ground distribution network interconnecting with base stations. The ground segment provides highly reliable and timely global connection between all CSEL ground elements utilizing the Defense Information System Network.

USER SEGMENT

8-27. The user segment equipment consists of—

- AN/PRC-7 radio set.
- J-6431/PRQ-7 radio set adapter (RSA) also referred to as the loader.
- Combat survivor evader locator planning computer (CPC).
- CPC program software.

AN/PRQ-7 Radio Set

8-28. The AN/PRQ-7 (refer to Figure 8-5) provides data communications geo-positioning, voice beacons. The RSA provides the physical interface the CPC and two operational AN/PRQ-7s. One AN/PRQ-7 serves as the reference in the RSA to acquire and store GPS almanac, ephemeris and time for the transfer to the other (target) AN/PRQ-7. The CPC host CSEL application software that allows loading of the AN/PRQ-7 through the RSA. A window operating environment is used to load a target AN/PRQ-7 with mission specific data and transfer GPS key loading. Loading current almanac and ephemeris data speed the satellite acquisition process in the GPS receiver. Transfer of current GPS data speeds the calculation of user position and transfer of current time allows faster acquisition of GPS.

Figure 8-5. AN/PRQ-7 radio set

8-29. The AN/PRQ-7 radio set has the following capabilities and characteristics—
- Water resistant.
- GPS receiver.
- Secure data UHF SATCOM transmit and receive capability.
- VHF/UHF voice and beacon.
- Low probability of exploitation of one way transmission.
- Search and rescue satellite transmission.

AN/PRC-90-2 TRANSCEIVER

8-30. The AN/PRC-90-2 is a LOS dual channel, personal survival transceiver used primarily used for communications between a downed crewman and a rescue aircraft. It has two preselected frequencies for voice and beacon transmissions. The signal is not secure and can be easily intercepted leaving isolated personnel limited to short voice transmissions.

8-31. The AN/PRC-90-2 (refer to Figure 8-6) can transmit a beacon (attention getting warble tone) on 243.0 MHz, voice on 243.0 or 282.2 MHz and Morse Code in modulated continuous wave (CW) mode on 243.0 MHz. It also has the capability of receiving voice communications on 243.0 and 282.2 MHz. The distance for LOS transmission also depends on conditions such as weather, terrain and battery power. (Refer to TM 11-5820-1049-12 for more information on the AN/PRC-90-2.)

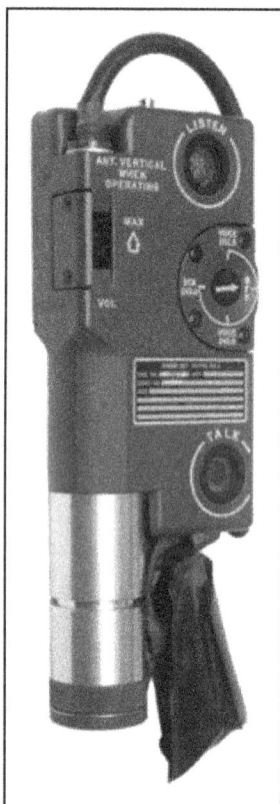

Figure 8-6. AN/PRC-90-2 transceiver

AN/PRC-112 COMBAT SEARCH AND RESCUE TRANSCEIVER

8-32. The AN/PRC-112 combat search and rescue transceiver is a replacement for the AN/PRC-90-2. The AN/PRC-112 has frequency ranges of—

- AM voice on 121.5 MHz, 243 MHz and 282.8 MHz.
- UHF frequency of 225–320 MHz.

8-33. The AN/PRC-112 (refer to Figure 8-7) operates in the following modes: voice, beacon, transponder mode, 406 search and rescue satellite, and UHF SATCOM. It is also dependant on the program loader KY-913 which has a keypad for data entry and an eight character display used to display the entered data and messages to the operator. The program loader attaches to the radio during programming and supplies the required power to the radio when attached. (Refer to TM 11-5820-1037-13&P for more information on the AN/PRC-112.)

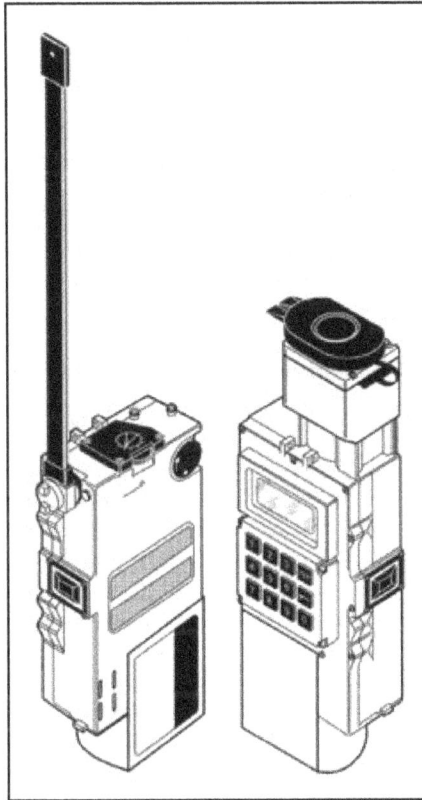

Figure 8-7. AN/PRC-112 and program loader KY-913

JOINT TACTICAL RADIO SYSTEM

8-34. The JTRS is the DOD radio of choice for radio requirements. The components of JTRS include airborne maritime fixed station, ground mobile radio, and handheld man-pack small form fit. JTRS are software based networking radios that will deliver networks to the mounted, dismounted, and un-mounted joint force.

> *Note.* At the time of publication JTRS had not been fielded to Army units and was in the process of being developed and tested. Pre-engineering design model ground mobile radios were available in the Experimental BCT Future Combat System.

8-35. The concept behind the JTRS family of radios (refer to Figure 8-8 for an example of the JTRS) is for all military services to migrate toward a commonality of media among Soldiers, while concurrently out-pacing the growth rate of information exchange requirements and eventually realizing a fully digitized tactical environment. JTRS lays the foundation for achieving network connectivity across the RF spectrum. The network will provide the means for low-to-high rate digital information exchange, both vertically and horizontally, between warfighting elements. It will also enable connectivity to civil and national authorities.

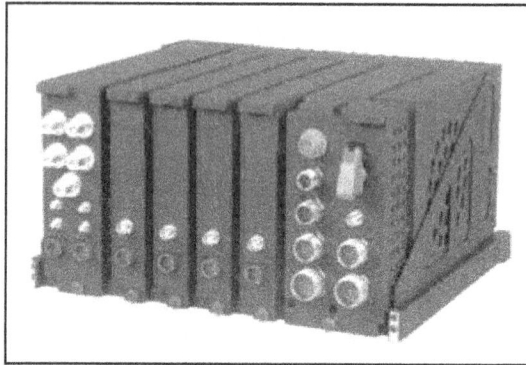

Figure 8-8. Joint tactical radio system ground mobile radio

8-36. The JTRS was designed to meet the emerging service needs for secure, multiband/multimode, high capacity digital radios for the future tactical environments. The JTRS provides increased interoperability among the Services, reduce upgrade costs through software programming (add new capabilities, change wave forms, and provide waveform enhancements), and support future legacy communications requirements.

8-37. The JTRS has ease of operation, redundancy, and security. It also has network capable, demand adaptive (dynamic bandwidth management), reliable, maintainable, deployable, and more survivable than the current generation of analog radios and stovepipe networks. The key features of the JTRS family of radios systems include—

- Simultaneous multichannel operation; has a fixed radio requirement for a minimum of four-channel operation (threshold) scalable to 10 channels (objective).
- Narrowband and wideband waveforms currently used in the 2 MHz and 2 GHz frequency range, to include HF ALE, SINCGARS, VHF AM 8.88 kHz operation for European ATC, ATC data links, HAVEQUICK I/II, UHF SATCOM DAMA, EPLRS, and Link 16.
- Increased throughput for data communications capabilities, including commercial waveforms.
- Multimode support for voice, data, video, and other communications.
- Integrated GPS, information security, modem, and baseband processing functions.
- Provide networking such as cross banding, bridging, relay, IP compatibility, and near real-time task organization.
- New capabilities, or provides waveform upgrades, as required.
- Extend with modular hardware and software, and can be reconfigured in the tactical environment.
- Interface with inventory PAs, antennas, and ancillary equipment.
- Operations in various domains—airborne, maritime/fixed, vehicular, dismounted (manpack), and handheld.

8-38. JTRS radios range from low cost terminals with limited waveform support, to multiband, multimode, and multichannel radios supporting advanced narrowband and wideband waveform capabilities with integrated computer networking features. The JTRS family will be open system architecture, interoperable with current legacy communications systems, capable of future technology insertion, and capable of providing both LOS, and BLOS, communications capabilities to the Soldier.

8-39. JTRS has several functionalities, it is—

- **To the user**—plug and play voice, high data throughput, and video-capable communications in a transparent network, with the ability to expand and modify the capacity and capability of the individual radio, links, and networks to accommodate user demands.
- **To the communicator**—intensive planning, management, and control:
  - Automated central planning and management; distributed technical control.
  - Information security, spectral efficiency, and electromagnetic interference (EMI)/electromagnetic compatibility.
  - Gateways to other systems (military, civil, joint, multinational, host, and nation).
- **The joint tactical radio family of systems,** which are scalable hardware configurations and multiple programmable waveforms and modes, capable of being operated and monitored while unattended, and remotely controlled and have standard interfaces and legacy radio emulation to operate in selected legacy radio nets.
- **The joint tactical Internet, to include—**
  - All hardware and software to form and manage a seamless mobile tactical radio Internet.
  - Common operating environment and dynamic power management.
  - Dynamic routing and traffic load management.
  - Embedded position location and automatic SA feed to the network.

JTRS WAVEFORM

8-40. Wireless tactical networking is one of the most critical capabilities a JTRS software defined radio will provide to the Soldier. The JTRS networking waveforms enable extension of networking to the battalion, company, and dismounted Soldiers.

8-41. The initial increment of JTRS being developed includes three networking, one BLOS waveform, and ten new software defined radios. The new networking waveforms are Soldier Radio Waveform, focused on the disadvantaged user using size, weight, and power constrained radios; Wideband Networking Waveform, for use on more capable vehicular, rotary-wing and fixed-wing aircraft; and the Joint Aerial Network-Tactical Edge for the fast moving aerial fleet that requires a very low latency capability. The BLOS waveform is the Mobile User Objective System, which will provide more capacity and throughput than the current UHF SATCOM system.

8-42. Each of these waveforms fills a particular operational need in the tactical environment, yet each provides a common transport function for IP-based traffic. The reprogrammable nature of the radio allows selection of the software waveform giving it multiple radios and networking capabilities, including legacy capabilities in one joint tactical radio set.

8-43. The waveform software developed for JTRS includes not only the actual RF signal, but the entire set of radio functions that occur from the user input to the RF output and vice versa. For example, in the transmitting JTRS, the waveform software will control the receipt of the data (either analog or digital) from the input device and manage the encoding. The encoded data is passed to the encryption engine. The resultant encoded/encrypted data stream is modulated into an intermediate frequency signal. Finally, the intermediate frequency signal is converted into a RF signal and transmitted to the antenna. These same functions will be reversed in the receiving JTRS with the ultimate output of the data to the user.

JTRS RIFLEMAN RADIO

8-44. The Rifleman Radio (refer to Figure 8-9) will provide Soldiers vertical and horizontal intra-squad network connectivity to achieve the information dominance deemed critical to successfully conduct dismounted operations independent of any vehicle or other communications infrastructure.

Note. The Rifleman Radio capabilities are currently being tested. The radio is projected to be fielded in FY10.

8-45. The Rifleman Radio will enable the individual Soldier to operate in a tactical voice network with other team members, team and squad leaders via a networking waveform (i.e., Soldier Radio Waveform). It will provide controlled unclassified real-time intra-squad C2 voice communications and transmit position location information enabling—

- A squad to employ much bolder and more sophisticated tactics to attack identified threats decisively.
- Increased speed of movement when conducting individual movement techniques as part of Fire Team and Squad.
- Improved networked communications while dispersed in complex terrain.
- Increased speed of maneuver, a reduced risk of potential fratricide, increased flexibility to transition missions on the move, more bold and sophisticated tactics, and the ability to attack identified threats decisively.
- A reduced exposure to the enemy, synchronized fire and maneuver in complex terrain, increased team movement distances, and a reduced limitation on movement locations.
- Soldiers to communicate with leaders when out of visual contact and shouting distance to conduct movement techniques as part of a squad.
- Leaders to display individual position location information of squad members (via an external display device or as part of a Ground Soldier Ensemble) when out of visual contact to coordinate fire and maneuver.
- Improved SA for leaders to make informed and timely decisions.

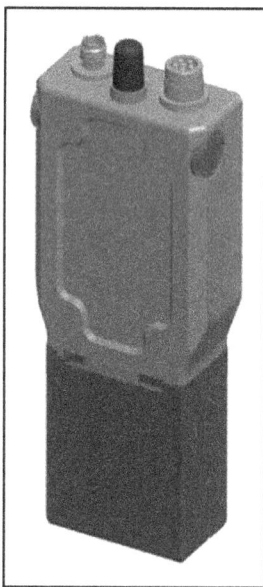

Figure 8-9. Rifleman radio

Chapter 9

Antennas

All radios, whether transmitting or receiving, require an antenna. This chapter addresses antenna fundamentals, concepts and terms, ground effects, antenna length, types of antennas, as well as examples of antenna field repairs.

ANTENNA FUNDEMENTALS

9-1. Simplex operation, or one-way-reversible, consists of sending and receiving radio signals on one antenna. It is normally used by SC radios. Two antennas are used during duplex operation: one for transmitting and one for receiving. In either case, the transmitter generates a radio signal; a transmission line delivers the signal from the transmitter to the antenna.

9-2. The transmitting antenna sends the radio signal into space toward the receiving antenna, which intercepts the signal and sends it through a transmission line to the receiver. The receiver processes the radio signal so it can either be heard or used to operate a data device, such as an AN/UXC-10 facsimile. Figure 9-1 is an example of a typical transmitter and receiver connection.

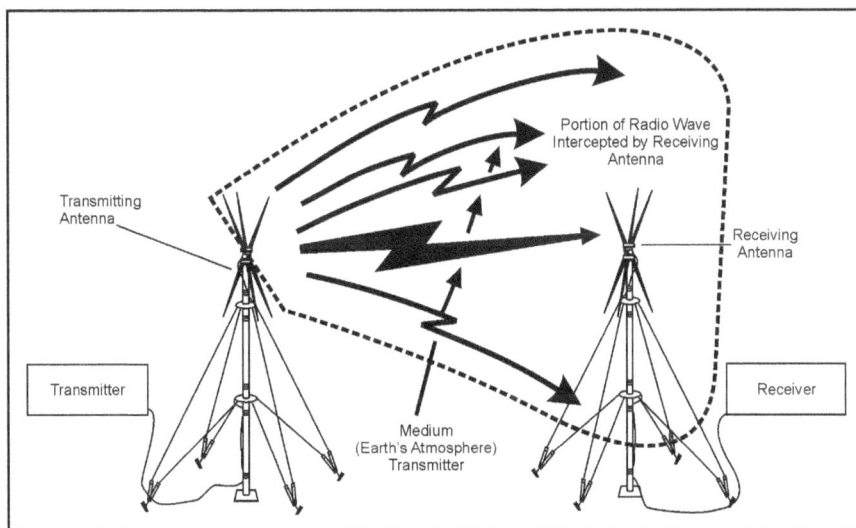

Figure 9-1. A typical transmitter and receiver connection

9-3. The function of an antenna depends on whether it is transmitting or receiving. A transmitting antenna transforms the output RF signal, in the form of an alternating electrical current produced by a radio transmitter (RF output power), into an electromagnetic field that is radiated through space; the transmitting antenna converts energy from one form to another form. The receiving antenna reverses this process; it transforms the electromagnetic field into electrical energy that is delivered to a radio receiver.

ANTENNA CONCEPTS AND TERMS

9-4. To select the right antenna, certain concepts and terms must be understood. The following paragraphs address several basic terms and relationships which help the reader understand antenna fundamentals.

FORMING A RADIO WAVE

9-5. When an alternating electric current flows through a conductor (wire), electric and magnetic fields are created around the conductor. If the length of the conductor is very short compared to a wavelength, the electric and magnetic fields will generally die out within a distance of one or two wavelengths. However, as the conductor is lengthened, the intensity of the field enlarges. Thus, an ever increasing amount of energy escapes into space.

RADIATION

9-6. Once a wire is connected to a transmitter and properly grounded, it begins to oscillate electrically, causing the wave to convert nearly all of the transmitter power into an electromagnetic radio wave. The electromagnetic energy is created by the alternating flow of electrons impressed on the bottom end of the wire. The electrons travel upward on the wire to the top, where they have no place to go and they are bounced back toward the lower end. As the electrons reach the lower end in phase (for example, they are in step with the radio energy then being applied by the transmitter) the energy of their motion is strongly reinforced as they bounced back upward along the wire. This regenerative process sustains the oscillation. The wire is resonant at the frequency at which the source of energy is alternating.

9-7. The radio power supplied to a simple wire antenna appears nearly equally distributed throughout its length. The energy stored at any location along the wire is equal to the product of the voltage and the current at that point. If the voltage is high at a given point, the current must be low. If the current is high, the voltage must be low. The electric current reaches its maximum near the bottom end of the wire.

RADIATION FIELDS

9-8. When RF power is delivered to an antenna, two fields are created: an induction field, which is associated with the stored energy, and a radiation field. At the antenna, the intensities of these fields are large, and are proportional to the amount of RF power delivered to the antenna. At a short distance from the transmitting antenna, and traveling toward the receiving antenna, only the radiation field remains; this radiation field is composed of electric and magnetic components. Figure 9-2 is an example of the components of electromagnetic waves.

9-9. The electric and magnetic fields (components) radiated from an antenna form the electromagnetic field. It is responsible for transmitting and receiving electromagnetic energy through free space. A radio wave is a moving electromagnetic field that has velocity in the direction of travel. Its components are of electric and magnetic intensity arranged at right angles to each other.

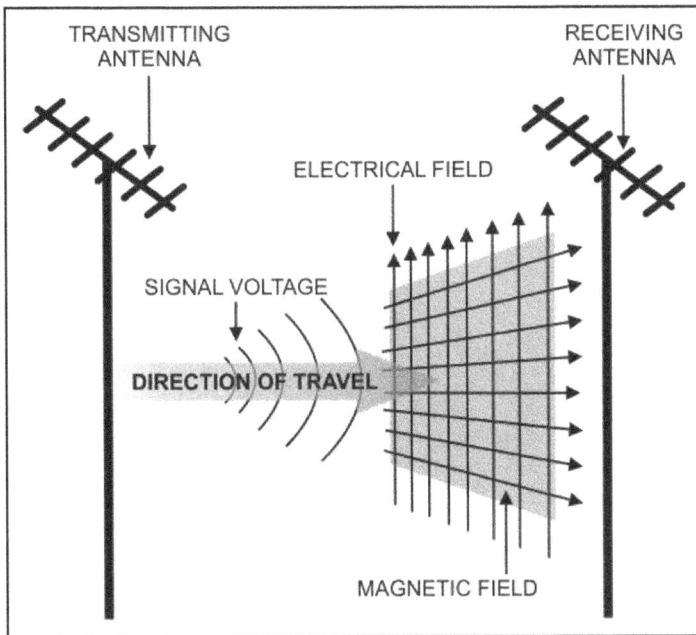

Figure 9-2. Components of electromagnetic waves

RADIATION PATTERNS

9-10. The radiation pattern is a graphical depiction of the relative field strength transmitted from, or received by, the antenna.

9-11. The full- or solid-radiation pattern is represented as a three-dimensional figure that looks somewhat like a doughnut with a transmitting antenna in the center. Figure 9-3 is an example of solid antenna radiation patterns. The top figure shows a quarter-wave vertical antenna; the middle figure shows a half-wave horizontal antenna, located one-half wavelength above the ground; and the bottom figure shows a vertical half rhombic antenna. (Omnidirectional and bidirectional antennas are discussed later in this chapter.)

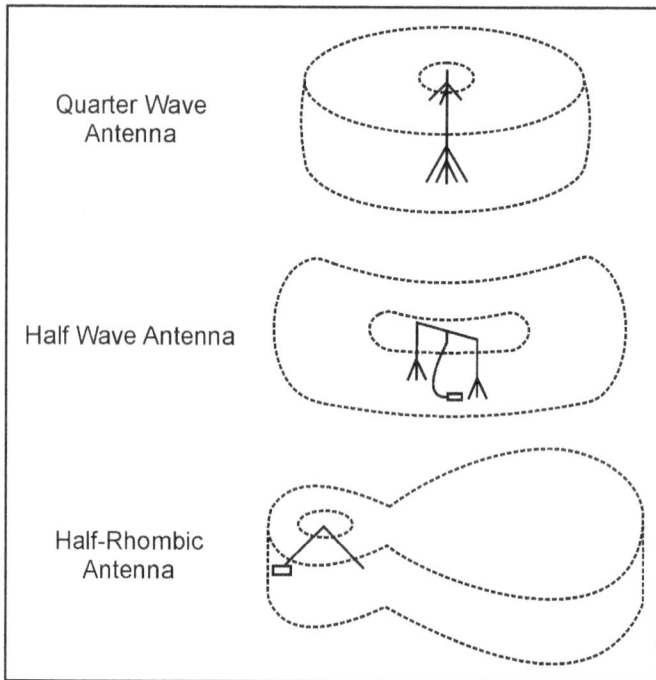

Figure 9-3. Solid radiation patterns

POLARIZATION

9-12. The polarization of a radiated wave is determined by the direction of the lines of force making up the electric field; polarization can be vertical, horizontal, or elliptical. When a single-wire antenna is used to extract (receive) energy from a passing radio wave, maximum pickup results if the antenna is oriented so that it lies in the same direction as the electric field component.

9-13. Horizontal or vertical polarization is satisfactory for VHF or UHF signals. The original polarization produced at the transmitting antenna is maintained as the wave travels to the receiving antenna. Therefore, if a horizontal antenna is used for transmitting, a horizontal antenna must be used for receiving.

Vertical Polarization

9-14. In a vertical polarized wave, the lines of electric force are at right angles to the surface of the earth. Figure 9-4 illustrates a vertical polarized wave. A vertical antenna is used for efficient reception of vertically polarized waves.

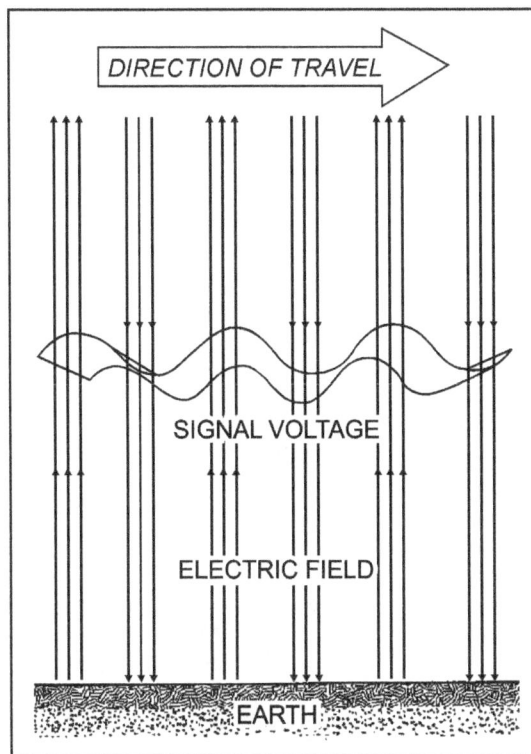

Figure 9-4. Vertically polarized wave

9-15. Vertical polarization is necessary at medium and low frequencies, because ground-wave transmission is used extensively. Vertical lines of force are perpendicular to the ground, and the radio wave can travel a considerable distance along the ground surface with a minimum amount of loss.

9-16. Vertical polarization provides a stronger received signal at frequencies up to approximately 50 MHz, when antenna heights are limited to 3.05 meters (10 ft) or less over land, as in a vehicular installation.

9-17. Vertically polarized radiation is less affected by reflections from aircraft flying over the transmission path. This factor is important in areas where aircraft traffic is heavy.

9-18. When vertical polarization is used, less interference is produced or picked up from strong VHF and UHF transmissions (television and FM broadcasts). This factor is important when an antenna must be located in an urban area that has television or FM broadcast stations.

Horizontal Polarization

9-19. In a horizontal polarized wave, the lines of electric force are parallel to the surface of the earth. A horizontal antenna is used for the reception of horizontally polarized waves. Figure 9-5 is an example of a horizontal polarized wave.

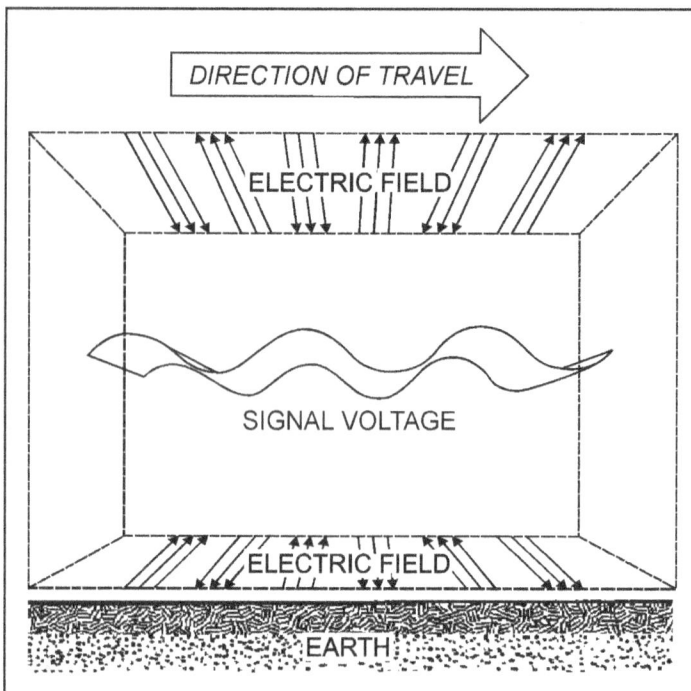

Figure 9-5. Horizontally polarized wave

9-20. At high frequencies, with sky wave transmission, it makes little difference whether horizontal or vertical polarization is used. The sky wave, after being reflected by the ionosphere, arrives at the receiving antenna elliptically polarized. Therefore, the transmitting and receiving antennas can be mounted either horizontally or vertically. However, horizontal antennas are preferred, since they can be made to radiate effectively at high angles and have inherent directional properties.

9-21. A simple horizontal, half-wave antenna is bidirectional. This characteristic is useful when minimizing interference from certain directions and masking signals from the enemy. Horizontal antennas are less likely to pick up man-made interference. When antennas are located near dense forests, horizontally polarized waves suffer lower losses, especially at frequencies above 100 MHz.

9-22. Small changes in antenna location do not cause large variations in the field intensity of horizontally polarized waves, when an antenna is located among trees or buildings.

Elliptical Polarization

9-23. In some cases, the field rotates as the waves travel through space. Under these conditions, both horizontal and vertical components of the field exist and the wave has elliptical polarization.

9-24. Satellites and satellite terminals use a type of elliptical polarization, called circular polarization. Circular polarization describes a wave whose plane of polarization rotates through 360 degrees as it progresses forward; the rotation can be clockwise or counterclockwise. Figure 9-6 is an example of a circular polarized wave. Circular polarization occurs when equal magnitudes of vertically and horizontally polarized waves are combined with a phase difference of 90 degrees. Depending on their phase relationship, this causes rotation either in one direction or the other.

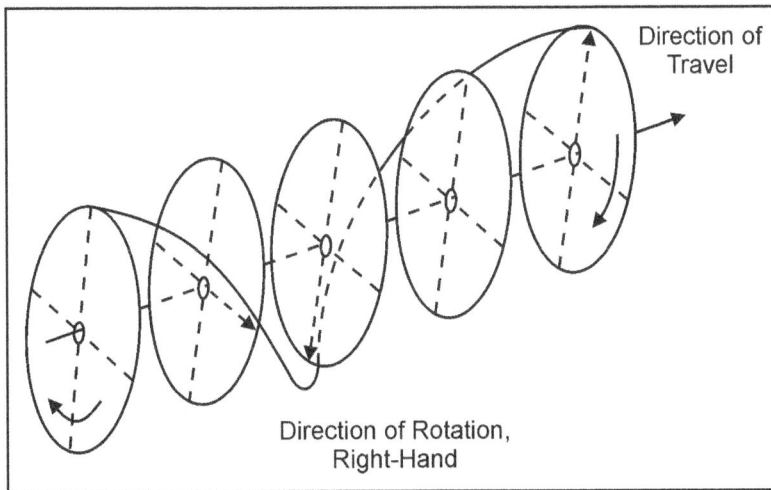

Figure 9-6. Circular polarized wave

DIRECTIONALITY

9-25. Vertical transmitting antennas radiate equally in horizontal directions; vertical receiving antennas accept radio signals equally from all horizontal directions. Thus, other stations operating on the same or nearby frequencies may interfere with the desired signal, making reception difficult or impossible. However, reception of a desired signal can be improved by using directional antennas.

9-26. Horizontal half-wave antennas accept radio signals from all directions. The strongest reception is received from a direction perpendicular to the antenna, while the weakest reception is received from the direction of the ends of the antenna. Interfering signals can be eliminated or reduced by changing the antenna installation, so that each end of the antenna points directly at the interfering station.

9-27. Communication over a radio circuit is satisfactory when the received signal is strong enough to override undesired signals and noise. Increasing the transmitting power between two radio stations increases communications effectiveness, as the receiver must be within range of the transmitter. Also, changing the types of transmission, changing to a frequency that is not readily absorbed or using a directional antenna aids in communications effectiveness.

RESONANCE

9-28. Antennas can be classified as either resonant or nonresonant, depending on their design. In a resonant antenna, almost all of the radio signals fed to the antenna are radiated. If the antenna is fed with a frequency other than the one for which it is resonant, much of the fed signal will be lost and will not be radiated. A resonant antenna will effectively radiate a radio signal for frequencies close to its design frequency. If a resonant antenna is used for a radio circuit, a separate antenna must be built for each frequency to be used on the radio circuit. A nonresonant antenna, on the other hand, will effectively radiate a broad range of frequencies with less efficiency. Resonant and nonresonant antennas are commonly used on tactical circuits. Resonance can be achieved in two ways: physically matching the length of the antenna to the wavelength and electronically matching the length of the antenna to the wavelength.

RECEPTION

9-29. The radio waves that leave the transmitting antenna will have an influence on and will be influenced by any electrons in their path. For example, as a HF wave enters the ionosphere, it is reflected or refracted back to the Earth by the action of free electrons in this region of the atmosphere. When the radio wave encounters the wire or metallic conductors of the receiving antenna, the radio wave's electric field will cause the electrons in the antenna to oscillate back and forth in step with the wave as it passes. The movement of these electrons within the antenna is the small alternating electrical current which is detected by the radio receiver.

9-30. When radio waves encounter electrons which are free to move under the influence of the wave's electric field, the free electrons oscillate in sympathy with the wave. This generates electric current which then creates waves of its own. These new waves are reflected or scattered waves. This process is electromagnetic scattering. All materials that are good electric conductors reflect or scatter RF energy. Since a receiving antenna is a good conductor, it too acts as a scatter. Only a portion of the energy which comes in contact with the antenna is converted into a received electrical power: a sizeable portion of the total power is re-radiated by the wire.

9-31. If an antenna is located within a congested urban environment or within a building, there are many objects that will scatter or reradiate the energy in a manner that can be detrimental to reception. For example, the electric wiring inside a building can strongly reradiate RF energy. If a receiving antenna is in close proximity to wires, it is possible for the reflected energy to cancel the energy received directly from the desired signal path. When this condition exists, the receiving antenna should be moved to another location within the room where the reflected and direct signals may reinforce rather than cancel each other.

Note. For more information on wave propagation refer to Training Circular 9-64.

RECIPROCITY

9-32. Reciprocity refers to the various properties of an antenna that apply equally, regardless of whether the antenna is used for transmitting or receiving. For example, the more efficient a certain antenna is for transmitting, the more efficient it will be for receiving the same frequency. The directive properties of a given antenna will be the same whether it is used for transmission or reception.

9-33. There is a minimum amount of radiation along the axis of the antenna. If this same antenna is used as a receiving antenna, it receives best in the same directions in which it produces maximum radiation (at right angles to the axis of the antenna). There is a minimum amount of signal received from transmitters located in the line with the antenna wire.

IMPEDANCE

9-34. Impedance is the relationship between voltage and current at any point in an alternating current circuit. The impedance of an antenna is equal to the ratio of the voltage to the current at the point on the antenna where the feed is connected (feed point). If the feed point is located at a point of maximum voltage point, the impedance is as much as 500 to 10,000 ohms.

9-35. The input impedance of an antenna depends on the conductivity or impedance of the ground. For, example, if the ground is a simple stake driven about a meter (3.2 ft) into earth of average conductivity, the impedance of the monopole may be double or even triple the quoted values. Because this additional resistance occurs at a point on the antenna circuit where the current is high, a large amount of transmitter power will dissipate as heat into the ground rather than radiated as intended. Therefore, it is essential to provide as good a ground or artificial ground (counterpoise) connection as possible when using a vertical whip or monopole.

9-36. The amount of power an antenna radiates depends on the amount of current which flows in it. Maximum power is radiated when there is maximum current flowing. Maximum current flows when the impedance is minimized which is when the antenna is resonated so that its impedance is pure resistance. (When capacitive reactance is made equal to inductive reactance, they cancel each other, and impedance equals pure resistance.)

BANDWIDTH

9-37. The bandwidth of an antenna is the frequency range over which it will perform within certain specified limits. These limits are with respect to impedance match, gain, and/or radiation pattern characteristics.

9-38. In the radio communication process, intelligence changes from speech or writing to low frequency signal that is used to modulate, or cause change, in a much higher frequency radio signal. When transmitted by an antenna, where it is picked up and reconverted into the original speech or writing. There are natural laws which limit the amount of intelligence or signal that can be transmitted and received at a given time. The more words per minute, the higher the rate of modulation frequency, so a wider or greater bandwidth is needed. To transmit and receive all the intelligence necessary, the antenna bandwidth must be as wide or wider that the signal bandwidth, otherwise it will limit the signal frequencies, causing voices and writing to be unintelligible. Too wide of a bandwidth is also bad, since it accepts extra voices and will degrade the S/N ratio.

ANTENNA GAIN

9-39. The antenna gain depends on its design. Transmitting antennas are designed for high efficiency in radiating energy, and receiving antennas are designed for high efficiency in picking up (gaining) energy. On many radio circuits, transmission is required between a transmitter and only one receiving station. Directed energy is radiated in one direction because it is useful only in that direction. Directional receiving antennas increase the energy gain in the favored direction and reduce the reception of unwanted noise in signals from other directions. Transmitting and receiving antennas should have small energy losses and should be efficient as radiators and receptors.

9-40. For example, current omnidirectional antennas, when employed in forward combat areas, transmit and receive signals equally in all directions, and provide an equally strong signal to both adversary EW units, and friendly units.

TAKE-OFF ANGLE

9-41. The antenna's take-off angle is the angle above the horizon that an antenna radiates the largest amount of energy (refer to Figure 9-7 for an example of an antenna take-off angle). VHF communications antennas are designed so that the energy is radiated parallel to the Earth (do not confuse take-off angle and polarization). The take-off angle of an HF communications antenna can determine whether a circuit is successful or not. HF sky wave antennas are designed for specific take-off angles, depending on the circuit distance. High take-off angles are used for short-range communications and low take-off angles are used for long range communications.

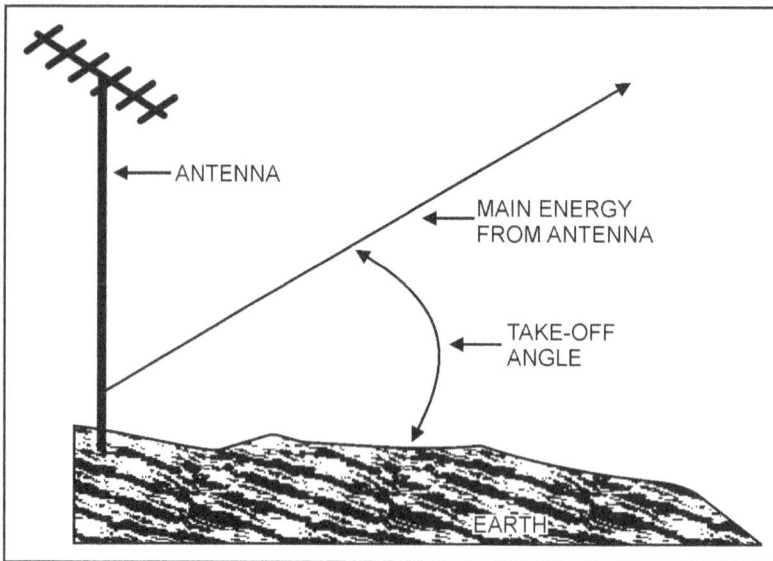

Figure 9-7. Antenna take-off angle

GROUND EFFECTS

9-42. Since most tactical antennas are erected over the earth, and not out in free space (except for those on satellites), the ground alters the free space radiation patterns of antennas. The ground will also have an effect on some of the electrical characteristics of antennas, specifically those mounted relatively close to the ground in terms of wavelength. For example, medium and HF antennas, elevated above the ground by only a fraction of a wavelength, will have radiation patterns that are quite different from the free-space patterns.

GROUNDED ANTENNA THEORY

9-43. When grounded antennas are used, it is important that the ground has as high conductivity as possible. This reduces ground loss, and provides the best possible reflecting surface for the down-going radiated energy from the antenna.

9-44. The ground is a good conductor for medium and low frequencies, and acts as a large mirror for the radiated energy. This results in the ground reflecting a large amount of energy that is radiated downward from an antenna mounted over it. Thus, a quarter-wave antenna erected vertically, with its lower end connected electrically to the ground, behaves like a half-wave antenna. Figure 9-8 is an example of a quarter-wave connected to the ground. Under these conditions, the vertical antenna (quarter wavelength) and the ground create the half wavelength. The ground portrays the quarter wavelength of radiated energy that is reflected to complete the half wavelength. At higher frequencies, artificial grounds constructed of large metal surfaces are common to provide better wave propagation.

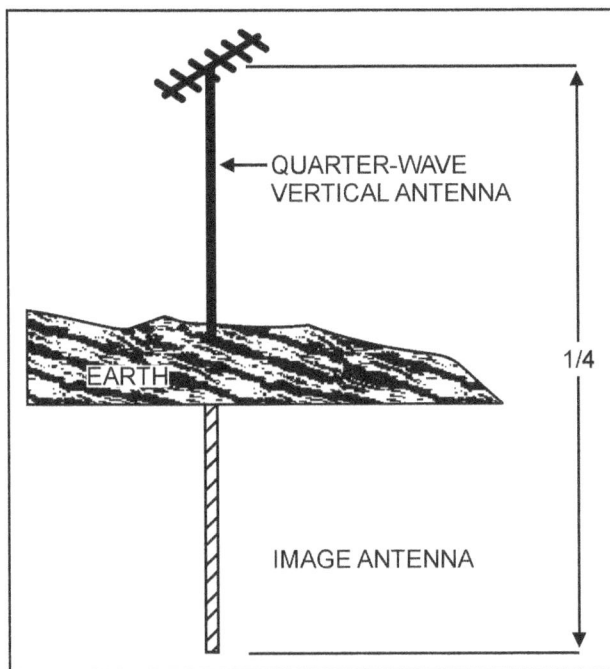

Figure 9-8. Quarter-wave antenna connected to ground

Types of Grounds

9-45. At low and medium frequencies, the Earth acts as a good conductor. The ground connection must be made in such a way as to introduce the least possible amount of resistance to ground. At higher frequencies, artificial grounds constructed of large metal surfaces are common.

9-46. The ground connections take many forms, depending on the type of installation and the loss that can be tolerated. In many simple field installations, the ground connection is made by one or more metal rods driven into the soil. Where more satisfactory arrangements cannot be made, ground leads can be connected to existing devices which are grounded. Metal structures or underground pipes systems are commonly used as ground connections. In an emergency, a ground connection can be made by forcing one or more bayonets into the soil.

Soil Conditions

9-47. When an antenna is erected over soil with low conductivity, treat the soil to reduce resistance. Soil ground conditions are categorized as favorable, less favorable, or unfavorable. The following paragraphs address a variety of grounding techniques that can be used during these soil conditions.

Favorable Soil Conditions

9-48. Ground connections take many forms, depending on the type of installation and the loss that can be tolerated. In many simple field installations, one or more metal rods driven into the soil make the ground connection. When more satisfactory arrangements cannot be made, ground leads can be connected to existing devices that are grounded. Metal structures or underground pipe systems are commonly used as ground connections. In an emergency, forcing one or more bayonets into the soil can make a ground connection.

Less Favorable Soil Conditions

9-49. When an antenna must be erected over soil with low conductivity, treat the soil with substances that are highly conductive when in solution, to reduce its resistance.

9-50. For simple installations, a single ground rod can be fabricated in the field from the pipe or conduit. It is important that a low resistance connection be made between the ground wire and the ground rod. The rod should be cleaned thoroughly by scraping and sand papering at the point where the connection is to be made, and a clean ground clamp should be installed. A ground wire can then be soldered or joined to the clamp; this joint should be covered with tape to prevent an increase in resistance because of oxidation.

Unfavorable Soil Conditions

9-51. When an actual ground connection cannot be used because of the high resistance of the soil, or because a large buried ground system is not practical, either a counterpoise or a ground screen may be used to replace the usual direct ground connection.

Couterpoise

9-52. When an actual ground connection cannot be used because of the high resistance of the soil or because a large buried ground system is not practical, a counterpoise may be used to replace the usual direct ground connection. The counterpoise consists of a device made of wire that is erected a short distance above the ground, and insulated from it. The size of the counterpoise should be equal to, or larger than, the size of the antenna. Figure 9-9 is an example of wire counterpoise.

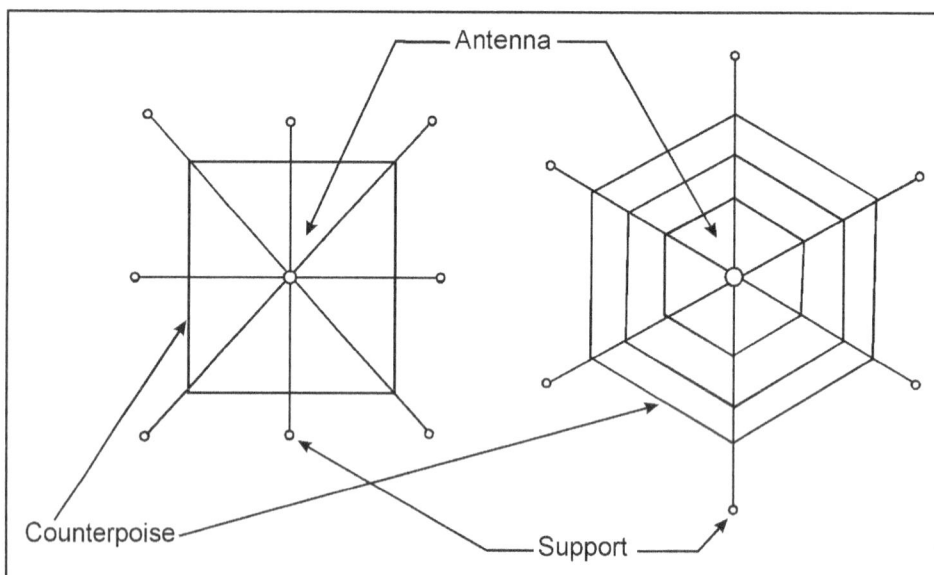

Figure 9-9. Wire counterpoise

9-53. When the antenna is mounted vertically, the counterpoise should be made into a simple geometric pattern; perfect symmetry is not required. The counterpoise appears to the antenna as an artificial ground that helps to produce the required radiation pattern.

9-54. In some VHF antenna installations on vehicles, the metal roof of the vehicle (or shelter) is used as a counterpoise for the antenna. Small counterpoises of metal mesh are sometimes used with special VHF antennas that must be located a considerable distance above the ground.

Ground Screen

9-55. A ground screen consists of a fairly large area of metal mesh or screen that is laid on the surface of the ground under the antenna. There are two specific advantages in using ground screens. First, the ground screen reduces ground absorption losses that occur when an antenna is erected over ground with poor conductivity. Second, the height of the antenna can be set accurately. Thus, the radiation resistance of the antenna can be determined more accurately.

ANTENNA LENGTH

9-56. The antenna has both a physical and electrical length; the two are never the same. The reduced velocities of the wave on the antenna, and a capacitive effect (known as end effect), make the antenna seem longer electrically than it is physically. The contributing factors are the ratio of the diameter of the antenna to its length, and the capacitive effect of terminal equipment (insulators, clamps) used to support the antenna.

9-57. To calculate the physical length of an antenna, use a correction of 0.95 for frequencies between 3.0–50.0 MHz. Table 9-1 provides antenna length calculations for a half-wave antenna.

Table 9-1. Antenna length calculations

The formula below calculates the half-wave length, and uses a correction of 0.95 for frequencies between 3 and 50 MHz. The same formula calculates the height above ground for HF wire antennas.		
Length (meters)	=150 X 0.95/frequency in MHz	=142.5/frequency in MHz
Length (ft)	=492 X 0.95/frequency in MHz	=468/frequency in MHz
The length of a long wire antenna (one wavelength or longer) for harmonic operation is calculated by using the following formula:		
Length (meters)	=150 X (N-0.05)/frequency in MHz	
Length (ft)	=492 X (N-0.05)/frequency in MHz	
Where N equals the number of half-wave lengths in the total length of the antenna. For example, if the number of half-wave lengths is 3 and the frequency in MHz is 7, then: Length (meters)=150(N-0.05)/frequency in MHz		
=150(3-0.05)/7	=150 X 2.95/7	=63.2 meters
Note. For HF antennas: a half wavelength in meters is 143/f where *f* is the frequency in MHz. If the frequency is 30 MHz, the wavelength is 5 meters. Often a half wavelength dipole is used and is center fed.		

ANTENNA ORIENTATION

9-58. The orientation of an antenna is extremely important. Determining the position of an antenna in relation to the points of the compass can make the difference between a marginal and good radio circuit.

Azimuth

9-59. If the azimuth of the radio path is not provided, the azimuth should be determined by the best available means. The accuracy required in determining the azimuth of the path depends on the radiation pattern of the directional antenna.

9-60. If the antenna beam width is very wide (for example, a 90 degree angle between half-power points), an error of 10 degrees in azimuth is of little consequence. However, in transportable operation, the rhombic and V antennas may have such a narrow beam as to require great accuracy in azimuth determination. The antenna should be erected for the correct azimuth; unless a line of known azimuth is available at the site, the direction of the path is best determined by a magnetic compass. Figure 9-10 is an example of a beam width measured on relative field strength and relative power patterns.

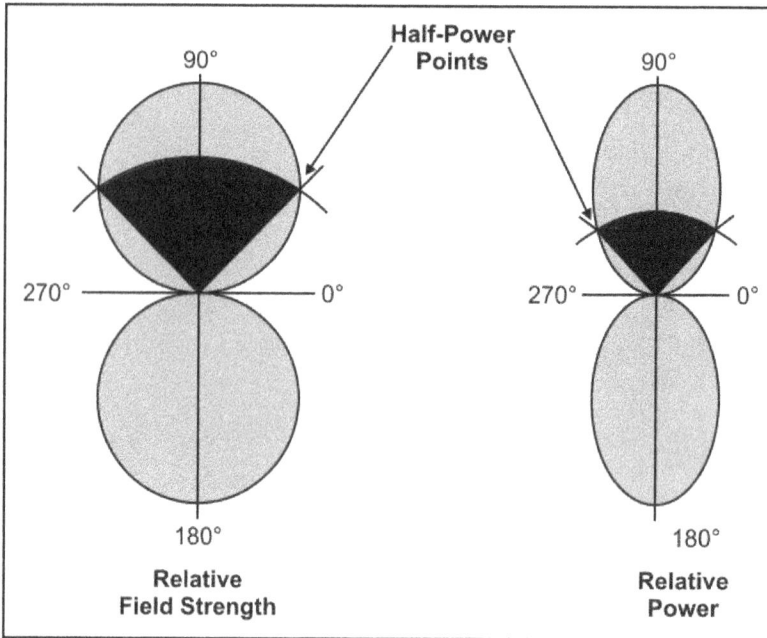

Figure 9-10. Beam width

9-61. Figure 9-11 is an example of a declination diagram. This example shows the relationship between the three north points (magnetic, grid and true) as represented on topographic maps by a declination diagram. It is important to understand the difference between the three and how to calculate from one to the other. Magnetic azimuths are determined by using magnetic instruments such as lensatic or M2 compasses while a grid azimuth is plotted on a map between two points, the points are joined together by a straight line and a protractor is used to measure the angle between grid north and drawn line. (Refer to FM 3-25.54 for more information on azimuths and map reading.)

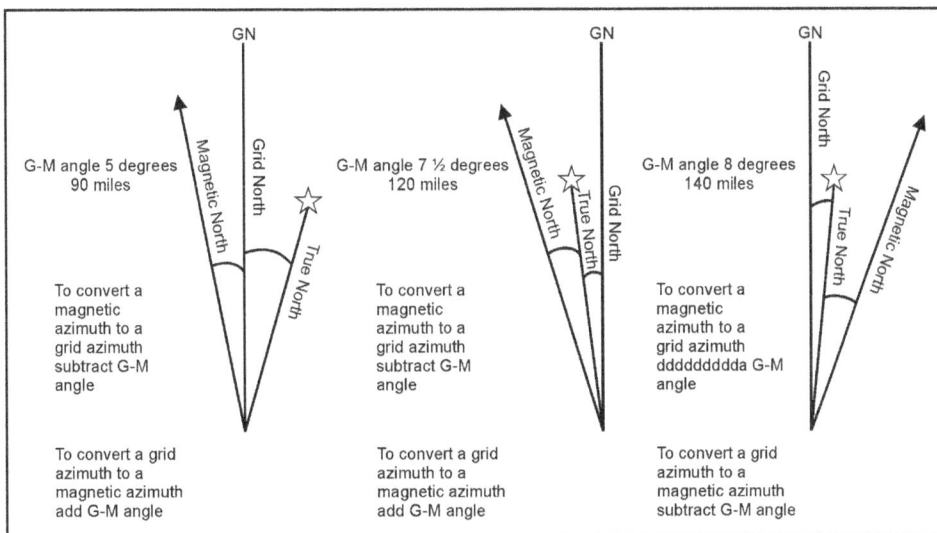

Figure 9-11. Example of a declination diagram

IMPROVEMENT OF MARGINAL COMMUNICATIONS

9-62. Under certain situations, it may not be possible to orient directional antennas to the correct azimuth of the desired radio path. As a result, marginal communications may suffer. To improve marginal communications—

- Check, tighten, and tape cable couplings and connections.
- Check to see that antennas are adjusted for the proper operating frequency (if possible).
- Change the heights of antennas.
- Move the antenna a short distance away, and in different locations, from its original location.
- Separate transmitters from receiving equipment, if possible.

9-63. An improvised antenna may change the performance of a radio set; use a distant station to test if an antenna is operating correctly. If the signal received from this station is strong, the antenna is operating satisfactorily. If the signal is weak, adjust the height and length of both the antenna and the transmission line, to receive the strongest signal at a given setting on the volume control of the receiver. This is the best method of tuning an antenna when transmission is dangerous or forbidden.

9-64. Impedance matching a load to its source is an important consideration in transmission systems. If the load and source are mismatched, part of the power is reflected back along the transmission line toward the source. This prevents maximum power transfer, and can be responsible for erroneous measurements of other parameters. It may also cause circuit damage in high-power applications.

9-65. The power reflected from the load interferes with the incident (forward) power, causing standing waves of voltages and current to exist along the line. Standing wave maximum-to-minimum ratio is directly related to the impedance mismatch of the load. Therefore, the standing wave ratio provides the means of determining impedance and mismatch.

TRANSMISSION AND RECEPTION OF STRONG SIGNALS

9-66. After an adequate site has been selected and the proper antenna orientation obtained, the signal level at the receiver will be proportional to the strength of the transmitted signal. If a high-gain antenna is used, a stronger signal can be obtained. Using a high quality transmission line (as short as possible and properly matched at both ends) can reduce losses between the antenna and the equipment.

```
┌─────────────────────────────────────────────────────────────────┐
│                         WARNING                                   │
│                                                                   │
│   Excessive signal strength may result in adversary intercept and │
│   interference, or in the operator interfering with adjacent      │
│   frequencies.                                                    │
│                                                                   │
└─────────────────────────────────────────────────────────────────┘
```

TYPES OF ANTENNAS

9-67. Tactical antennas are designed to be rugged; they permit mobility with the least possible sacrifice of efficiency. Some are mounted on the sides of vehicles that have to move over rough terrain; others are mounted on single masts, or suspended between sets of masts. All tactical antennas must be easy to install. Small antennas are mounted on the helmets of personnel who use the radio sets; large antennas must be easy to dismantle, pack, and transport.

9-68. A Hertz antenna (also known as a doublet, dipole, an ungrounded, or a half-wave antenna) can be mounted in a vertical, horizontal, or slanting position; it is generally used at higher frequencies (above 2 MHz). With Hertz antennas, the wavelength to which any wire electrically tunes depends directly upon its physical length. The basic Hertz antenna is center fed, and its total wire length is equal to approximately one half of the wavelength of the signal to be transmitted.

9-69. A Marconi antenna is a quarter-wave antenna with one end grounded (usually through the output of the transmitter or the coupling coil at the end of the feed line) which is required for the antenna to resonate. It is positioned perpendicular to the earth and is generally used at the lower frequencies. However, when used on vehicles or aircraft, Marconi antennas operate at high frequencies. In these cases, the aircraft or vehicle chassis becomes the effective ground for the antenna.

9-70. The main advantage of the Marconi antenna over the Hertz antenna is that, for any given frequency, the Marconi antenna is physically much shorter. This is particularly important in all field and vehicular radio installations. Typical Marconi antennas include the inverted L, and the whip.

9-71. The best kinds of wire for antennas are copper and aluminum. In an emergency, use any type that is available. The exact length of most antennas is critical. An expedient antenna should be the same length as the antenna it replaces.

HIGH FREQUENCY ANTENNAS

9-72. The following paragraphs describe HF NVIS communication and HF antennas. Refer to Appendix C for information on antenna selection.

Near-Vertical Incident Sky Wave Antenna, AS-2259/GR

9-73. The NVIS antenna, AS-2259/GR, is a lightweight sloping dipole omnidirectional antenna. Figure 9-12 is an example of the NVIS antenna. The NVIS is employed with HF radio communications in a 0–483 km (0 to 300 miles) range. It is capable of operating with older AM/HF radio sets, and was typically issued with the older IHFR.

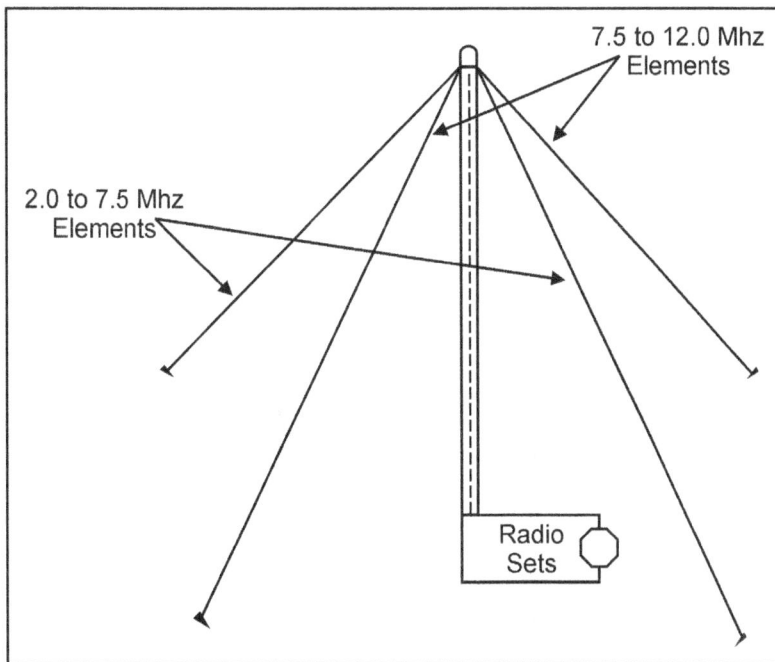

Figure 9-12. NVIS antenna, AS-2259/GR

Harris RF-1944, Inverted Vee HF Antenna

9-74. The Harris RF-1944 Inverted Vee antenna is a lightweight, broadband dipole COTS antenna that is primarily being fielded with the AN/PRC 150 (the older AS-2259/GR antenna is rarely used) The RF-1944 is primarily used because it is ideal for radios that have ALE and FH capabilities. The Harris RF-1944 antenna capabilities include—

- Horizontal polarization.
- Radiation patterns ideal for HF skywave communications from 0–500 miles (0–804.7 km).
- Bandwidth over the entire 1.6–30 MHz frequency range.
- Up to 20 watts power and 50 ohms input impedance.
- A gain of:
  - -16 dBi (gain in decibels) at 2 MHz.
  - -2 dBi at 30 MHz.
- Weight of less than four pounds.

9-75. The RF-1944 antenna does not include a mast. The primary components are a balun, two radiation elements with integral terminating loads, two ground stakes, a coaxial cable, a weighing throwing line, and a carrying bag. An added bonus for Soldiers is that the small, lightweight antenna can easily be carried in a rucksack.

Note. A balun is a device used to couple a balanced device or line to an unbalanced device or line.

V Antenna

9-76. The V antenna is a medium- to long-range, broadband sky wave antenna. It is used for point-to-point communications to ranges exceeding 4,000 km (2,500 miles). The V antenna consists of two wires arranged to form a V, with its ends at the apex (where the legs come together) attached to a transmission line (Figure 9-13). Radiation lobes off each wire combine to increase gain in the direction of an imaginary line bisecting the apex angle; the pattern is bidirectional. However, adding terminating resistors (300 ohms) to the far end of each leg will make the pattern unidirectional (in the direction away from the apex angle).

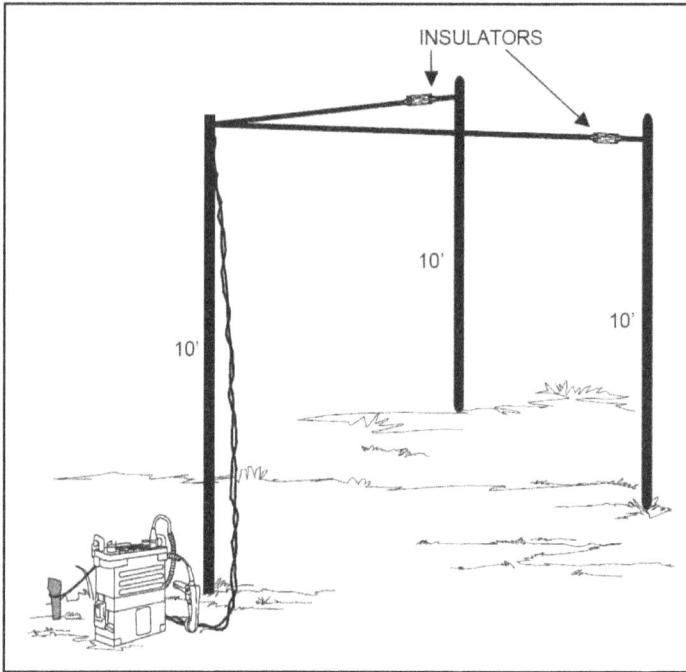

Figure 9-13. V antenna

9-77. The angle between the legs varies with the length of the legs to achieve maximum performance. Use Table 9-2 to determine the angle and the length of the legs. When the antenna is used with more than one frequency or wavelength, use an apex angle that is midway between the extreme angles determined by the chart.

Table 9-2. Leg angle for V antennas

Antenna Length (Wavelength)	Optimum Apex Angle (Degrees)
1	90
2	70
3	58
4	50
6	40
8	35
10	33

Vertical Half Rhombic Antenna and the Long Wire Antenna

9-78. The vertical half rhombic antenna and the long-wire antenna are two field expedient directional antennas. The long wire antenna directive pattern will radiate in both the horizontal and vertical planes and the vertical half rhombic antenna will radiate both to the front and back of the sloping wires if resistors are not used.

9-79. Figures 9-14 and 9-15 are examples of the vertical half rhombic antenna and the long wire antenna, respectively. These antennas consist of a single wire, preferably two or more wavelengths long, supported on poles at a height of 3–7 meters (10–20 ft) above the ground. However, the antennas will operate satisfactorily as low as 1 meter (approximately 3.2 ft) above the ground.

9-80. The far end of the wire is connected to the ground through a non-inductive resistor of 500–600 ohms. To ensure the resistor is not burned out by the output power of the transmitter, use a resistor rated at least one-half the wattage output of the transmitter. A reasonably good ground, such as a number of ground rods or a counterpoise, should be used at both ends of the antenna. The antennas are used primarily for transmitting or receiving HF signals.

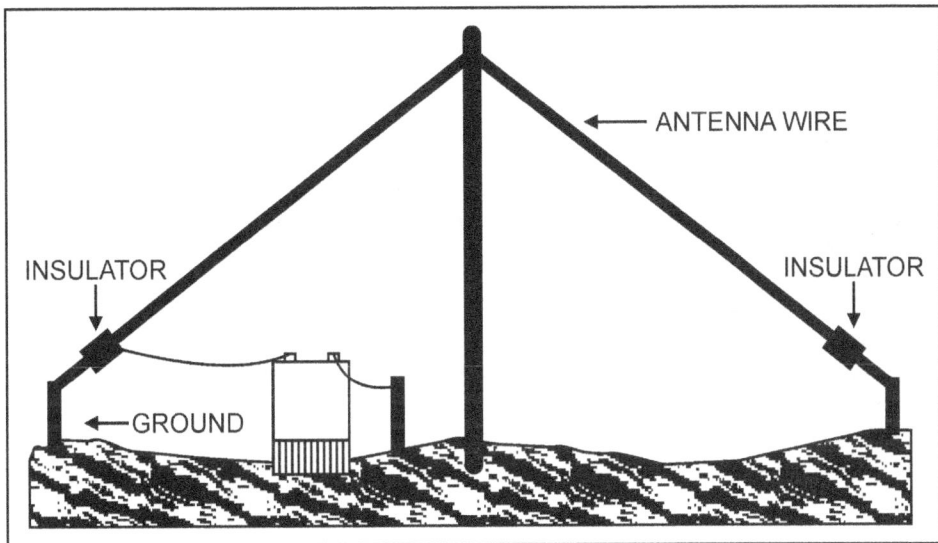

Figure 9-14. Vertical half rhombic antenna

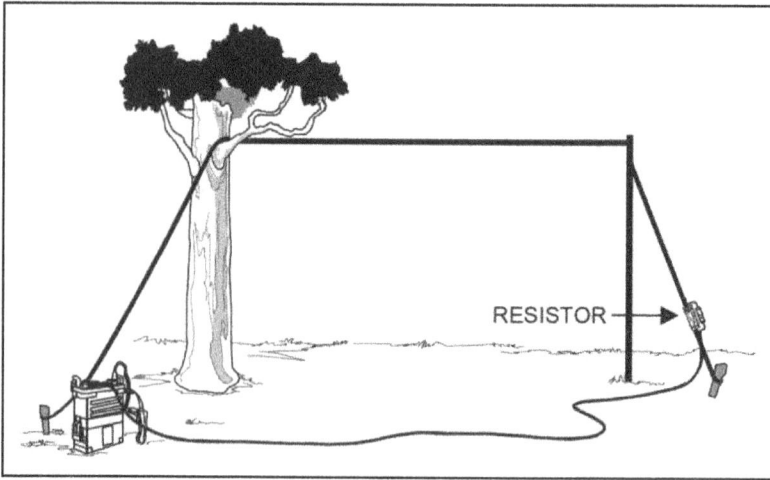

Figure 9-15. Long-wire antenna

Sloping V Antenna

9-81. The sloping V antenna is another field expedient directional antenna. To make construction easier, the legs may slope downward from the apex of the V (this is called a sloping V antenna). Figure 9-16 is an example of a sloping V antenna.

9-82. To make the antenna radiate in only one direction, add non-inductive terminating resistors from the end of each leg (not at the apex) to ground. The resistors should be approximately 500 ohms and have a power rating at least one half that of the output power of the transmitter being used. Without the resistors, the antenna radiates bi-directionally, both front and back. A balanced transmission line must feed the antenna.

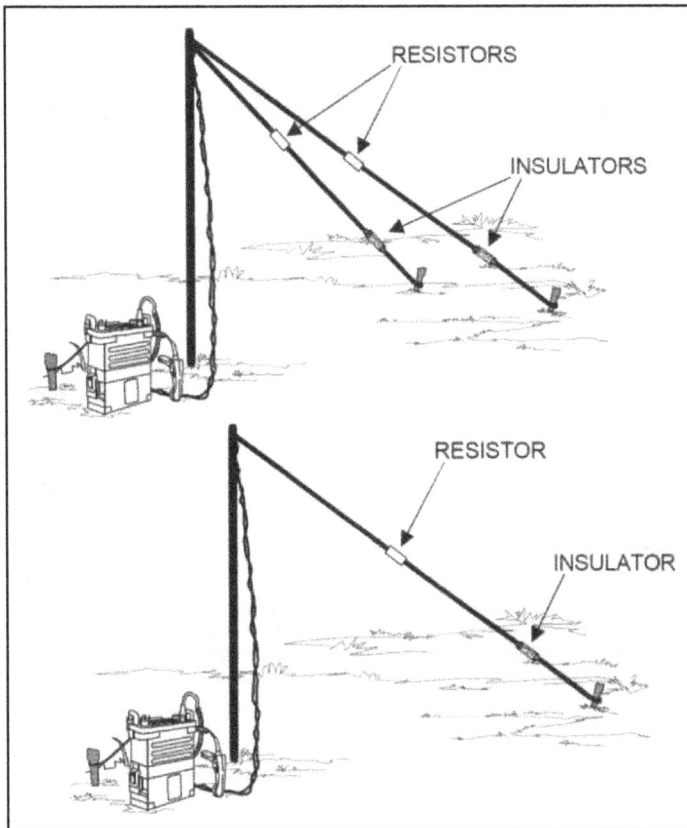

Figure 9-16. Sloping-V antenna

Inverted L Antenna

9-83. The inverted L is a combination antenna made up of vertical and horizontal wire sections. It provides omnidirectional radiation (when no resistors are being used) from the vertical element for ground wave propagation, and high-angle radiation from the horizontal element for short-range sky wave propagation, 0–400 km (0–250 miles). The classic inverted L has a quarter-wave vertical section and a half-wave horizontal section.

9-84. Table 9-3 outlines the frequency and the length of the horizontal element. Using a vertical height of 11–12 meters (35–40 ft), this combination will give reasonable performance for short-range sky wave circuits. Figure 9-17 is an example inverted L antenna.

Table 9-3. Frequency and inverted L
horizontal element length

Operating Frequency	Length of Horizontal Element
5.0–7.0 MHz	24.3 meters (80 ft)
3.5–6.0 MHz	30.4 meters (100 ft)
2.5–4.0 MHz	45.7 meters (150 ft)

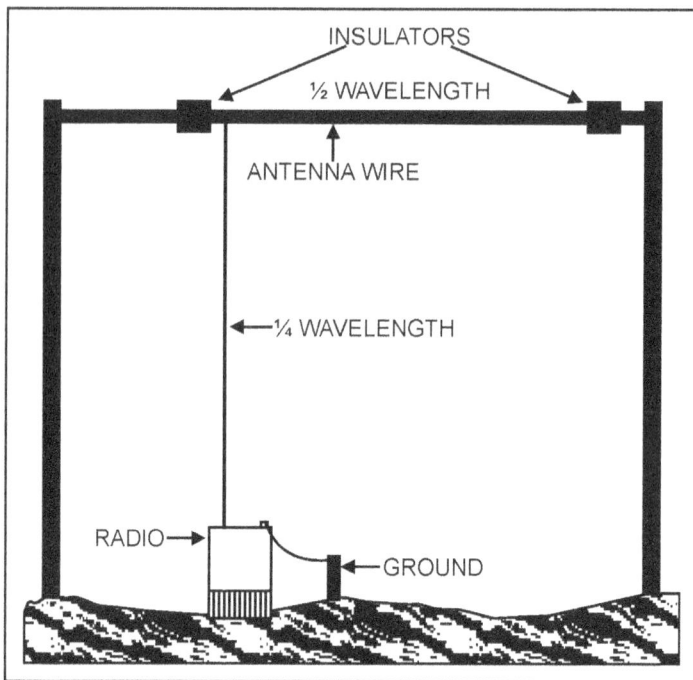

Figure 9-17. Inverted L antenna

Near-Vertical Incident Sky Wave Communications

9-85. The standard communications techniques used in the past will not support the widely deployed and fast moving formations of today's Army. Coupling this with the problems that can be expected in deploying multi-channel LOS systems with relays to keep up with present and future operation, HF radio and the NVIS mode take on new importance. The HF radio is quickly deployable, securable, and capable of data transmission. HF (such as the AN/PRC-150 [C]) will be the first, and frequently the only, means of communicating with fast-moving or widely separated units. With this reliance on HF radio, communications planners, commanders, and operators must be familiar with NVIS techniques and their applications and shortcomings in order to provide more reliable communications.

9-86. NVIS propagation is simply sky wave propagation that uses antennas with high angle radiation and low operation frequencies. Just as the proper selection of antenna can increase the reliability of a long range circuit, the same holds true for short range communications.

9-87. NVIS propagation uses high take-off angle (60–90 degrees) antennas to radiate the signal almost straight up. The signal is then reflected back from the ionosphere and returns to Earth in a circular pattern all around the transmitter. Because of near vertical radiation angle, there is no skip zone (skip zone is the area between the maximum ground wave distance and the shortest sky wave distances where no communications are possible). Communications are continuous out to several hundred kilometers from the transmitter. The nearly vertical angle of radiation also means lower frequencies must be used.

9-88. Generally, NVIS propagation uses frequencies up to 8 MHz. The steep up and down propagation of the signal gives the RTO the ability to communicate over nearby ridge lines, mountains, and dense vegetation. A valley location may give the RTO terrain shielding from hostile intercept or protect the circuit from ground wave and long wave interference. Antennas used for NVIS propagation need high take-off angle radiation with very little ground wave radiation. Refer to Figure 9-18 for an example of NVIS propagation.

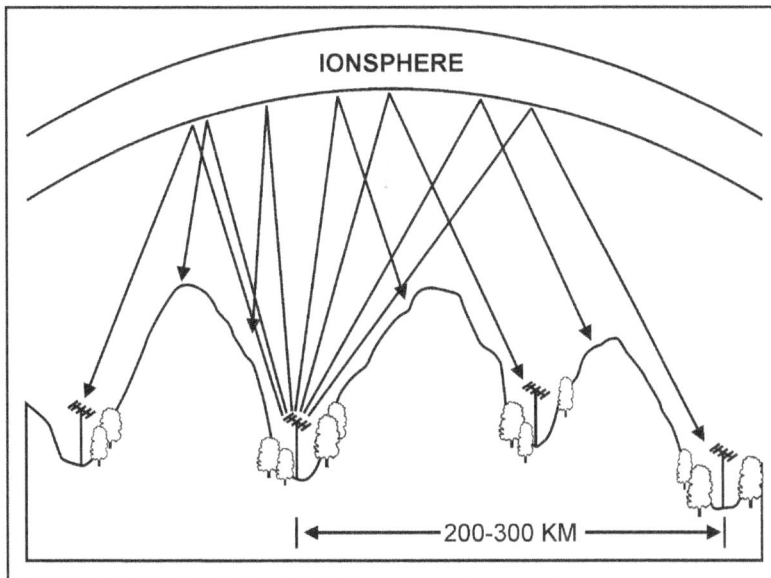

Figure 9-18. NVIS propagation

9-89. Using the HF antenna table matrix in Appendix C, the AS-2259/GR and the half wave dipole are the only antennas listed that meet the requirements of NVIS propagation. While the inverted V and inverted L have high angle radiation, they can also have strong ground wave radiation that could interfere with the close-in NVIS communications.

Disadvantages of Using the NVIS Concept

9-90. It is also important to understand that where both NVIS and ground wave signals are present, the ground wave can cause destructive interference. Proper antenna selection will suppress ground wave radiation and minimize this effect while maximizing the amount of energy going into the NVIS mode.

Advantages of Using the NVIS Concept

9-91. The following are advantages of using NVIS in a tactical environment—

- There are skip-zone-free omnidirectional communications.
- Terrain does not affect loss of signal. This gives a more constant received signal level over the operational range instead of one which varies widely with distance.
- Operators are able to operate from protected, dug-in positions. Thus tactical commanders do not have to control the high ground for HF communications purposes.
- Orientation, such as, doublets and inverted antennas are not as critical.

9-92. The following are advantages of using NVIS in an EW environment—

- **There is a lower probability of geolocation.** NVIS energy is received from above at very steep angles, which makes direction finding (DF) from nearby (but beyond ground wave range) locations more difficult.
- **Communications are harder to jam.** Ground wave jammers are subject to path loss. Terrain features can be used to attenuate a ground wave jammer without degrading the desired communication path. The jamming signal will be attenuated by terrain, while the sky wave

NVIS path loss will be constant. This will force the jammer to move very close to the target or put out more power. Either tactic makes jamming more difficult.

- **Operators can use low power successfully.** The NVIS mode can be used successfully with very low power HF sets. This will result in much lower probabilities of LPI/D.

VHF/UHF ANTENNAS

9-93. The following paragraphs address VHF/UHF antennas and their characteristics and capabilities.

Whip Antenna

9-94. Whip antennas for VHF tactical radio sets are usually 4.5 meters (15 ft) long. A vehicular whip antenna in HF operations has a planning range of 400–4,000 km (250–2,500 miles).

9-95. Two whip antennas are used with lightweight portable FM radios; a 0.9 meter (2.9 ft) long semi-rigid steel tape antenna, and a 3 meter (9.8 ft) long multi-section whip antenna. These antennas are made shorter than a quarter wavelength to ensure they are kept at a practical length. (A quarter wavelength antenna for a 5.0 MHz radio would be over 14 meters/45.9 ft long.) An antenna tuning unit, either built into the radio set or supplied with it, compensates for the missing length of the antenna. The tuning unit varies the electrical length of the antenna to accommodate a range of frequencies.

9-96. Whip antennas are used with tactical radio sets because they radiate equally in all directions on the horizontal plane. Since stations in a radio net lie in random directions and change their positions frequently, the radiation pattern is ideal for tactical communications.

9-97. When a whip antenna is mounted on a vehicle, the metal of the vehicle affects the operation of the antenna. Thus, the direction in which the vehicle is facing may also affect transmission and reception, particularly of distant or weak signals.

9-98. At lower frequencies where wavelengths are longer, it is impractical to use resonant-length tactical antennas with portable radio equipment, especially with vehicle-mounted radio sets. Tactical whip antennas are electrically short, vertical, base loaded types, fed with a nonresonant coaxial cable of about 52 ohms impedance. Figure 9-19 is an example of a whip antenna.

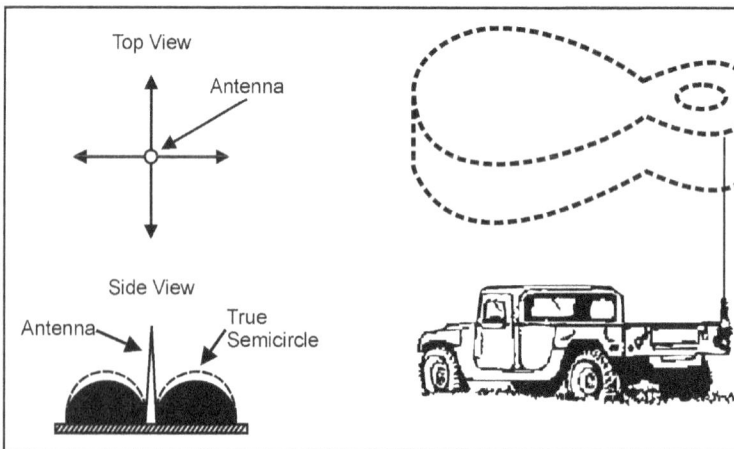

Figure 9-19. Whip antenna

9-99. To attain efficiency with a tactical whip, comparable to that of a half-wave antenna, the height of the vertical radiator should be a quarter wavelength. This is not always possible, so the loaded whip is used instead. The loading increases the electrical length of the vertical radiator to a quarter wavelength. The ground, counterpoise, or any conducting surface that is large enough, supplies the missing quarter-wavelength of the antenna.

9-100. A vehicle with a whip antenna mounted on the left rear side of the vehicle transmits its strongest signal in a line running from the antenna through the right front side of the vehicle. Similarly, an antenna mounted on the right rear side of the vehicle radiates its strongest signal in a direction toward the left front side. Figure 9-20 shows the best direction for whip antennas mounted on vehicles. The best reception is obtained from signals traveling in the direction shown by the dashed arrows on the figure.

9-101. In some cases, the best direction for transmission can be determined by driving the vehicle in a small circle until the best position is located. Normally, the best direction for receiving from a distant station is also the best direction for transmitting to that station.

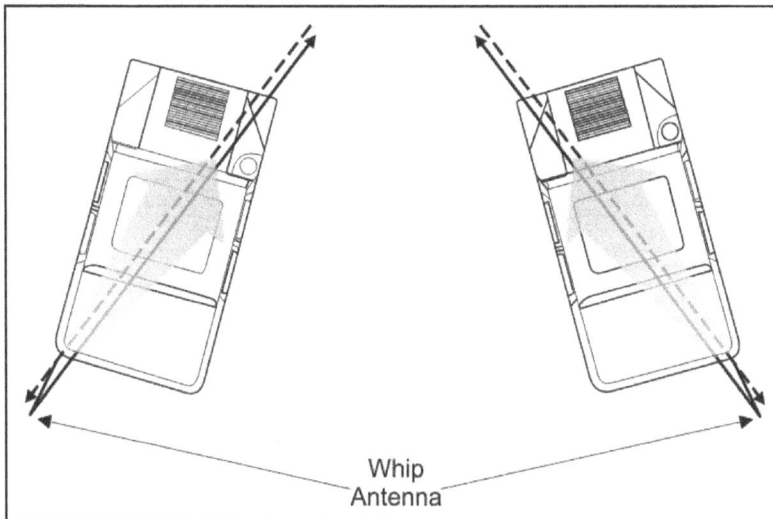

Figure 9-20. Whip antennas mounted on a vehicle

9-102. Sometimes, a whip antenna mounted on a vehicle must be left fully extended so that it can be used instantly while the vehicle is in motion. The base-mounted insulator of the whip is fitted with a coil spring attached to a mounting bracket on the vehicle. The spring base allows the vertical whip antenna to be tied down horizontally when the vehicle is in motion, and when driving under low bridges or obstructions. Even in the vertical position, if the antenna hits an obstruction, the whip usually will not break because the spring base absorbs most of the shock.

9-103. Some of the energy leaving a whip antenna travels downward and is reflected by the ground with practically no loss. To obtain greater distance in transmitting and receiving, it may be necessary to raise the whip antenna. However, when a whip antenna is raised, its efficiency decreases because it is further from the ground. Therefore, when using a whip antenna at the top of a mast, supply an elevated substitute for the ground (ground plane).

DANGER

When an antenna must be left fully extended while in motion, contact with overhead power lines must be avoided. Death or serious injury can result if a vehicular antenna strikes a high-voltage transmission line. If the antenna is tied down, be sure the tip protector is in place.

Broadband Omnidirectional Antenna

9-104.   The broadband omnidirectional, vertically polarized, VHF antenna system OE-254 (refer to Figure 9-21) is an improved tactical antenna. Table 9-4 shows planning ranges for the OE-254 antenna. The OE-254 antenna—

- Operates in the 30–88 MHz range without any physical adjustments.
- Has input impedance of 50 ohms unbalanced with an average voltage standing wave ratio (VSWR) of 3:1 or less, at RF power levels up to 350 watts.
- Is capable of being assembled and erected by one individual.
- Meets the broadband and power handling requirements of the frequency hopping multiplexer (FHMUX). (For more information on the OE-254 antenna refer to TM 11-5985-357-13.)

Table 9-4. OE-254 planning ranges

Terrain	High Power	Low Power (Nominal Conditions)
OE to OE		
Average Terrain	57.9 km (36 miles)	19.3 km (12 miles)
Difficult Terrain	48.3 km (30 miles)	
OE to Vehicle Whip		
Average Terrain	48.3 km (30 miles)	12.9 km (8 miles)
Difficult Terrain	40.3 km (25 miles)	

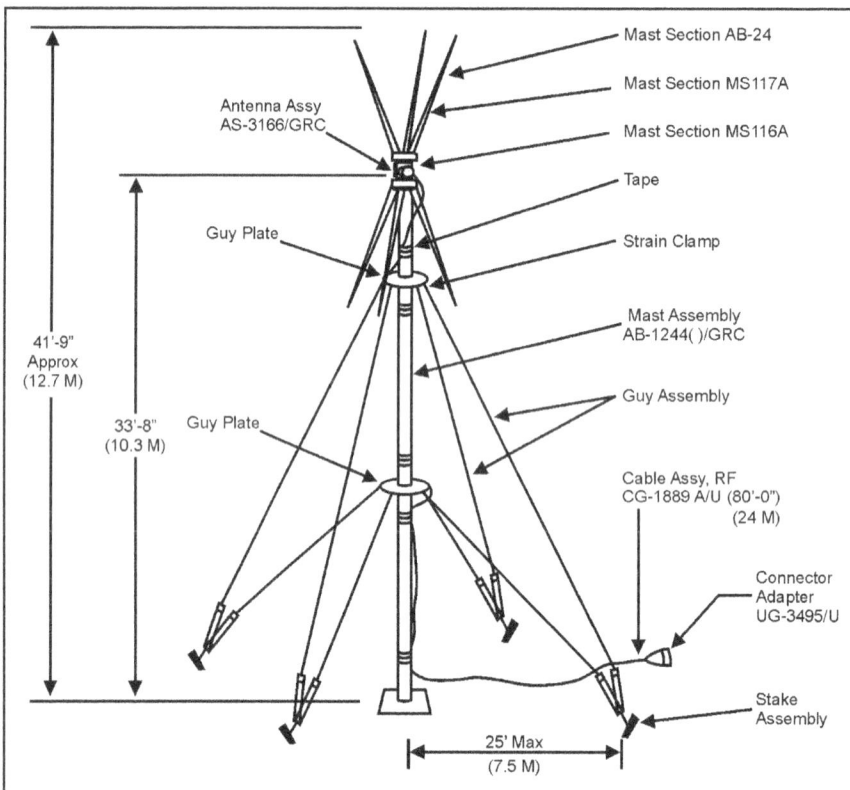

Figure 9-21. OE-254 broadband omnidirectional antenna system

Quick Erect Antenna Mast, AB 1386/U

9-105. The quick erect antenna mast (QEAM) is used for elevating tactical communications antennas to a maximum height of 33 ft (10 meters) which results in more reliable communications over extended ranges. The QEAM uses the same antenna elements and RF cable as the OE-254 antenna The QEAM will mount the OE-254, MSE and EPLRS antenna.

9-106. The mast can be deployed and operated in a ground or vehicular (wheeled and tracked) mounted configuration. It can also be erected in 7 ½ minutes by two Soldiers and only 15 minutes by one. Refer to Figure 9-22 for an example of the QEAM.

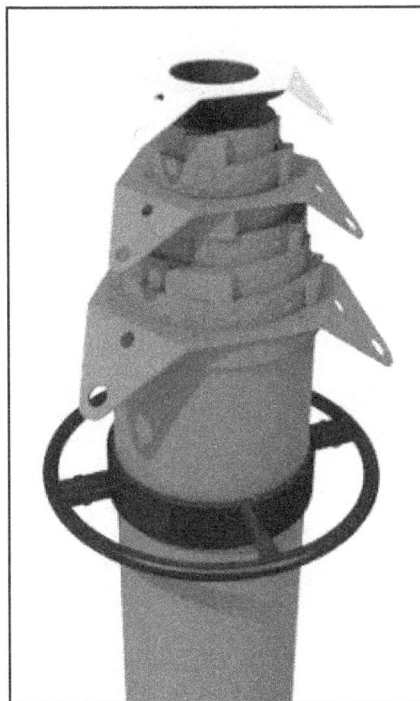

Figure 9-22. QEAM AB 1386/U

COM 201B Antenna

9-107. The COM 201B antenna is a commercial (from Atlantic Microwave Corporation) VHF/UHF vertically polarized, omnidirectional antenna that has become popular due to its versatility and unique design. The antenna was originally used by the USMC and is now a standard USMC item. It has a tripod leg structure that allows the antenna to be mounted directly on the ground or in a standard communications mast and can be quickly assembled and disassembles for transport and storage which makes it ideal in situations where there is not enough time to erect the OE-254. Refer to Figure 9-23 for an example of the COM 201-B.

Note. The COM 201B is not an Army issued replacement for the OE-254 antenna.

Figure 9-23. COM-201B antenna

9-108.   The antennas ease of operations makes it ideal for a field expedient antenna or mounting to a vehicle if more elevation is needed. The eye fitting at the top of the antenna facilitates suspending it from buildings or trees when a mast isn't available but more height is desired.

9-109.   The COM 201B antenna has the following characteristics and capabilities—

* Operates in the 30–88 MHz range.
* Vertically polarized.
* Input impedance of 50 ohms unbalanced with an average VSWR of 3:1 or less, at RF power levels up to 200 watts.
* Maximum power is directed towards the horizon with a typical antenna gain of +2 dB relative to an isotropic source.
* One individual can assemble and erect.
* Assembly can be stored in a space less than 36 inches by 10 inches in diameter.

OE-303, VHF Half Rhombic Antenna

9-110. The VHF half rhombic antenna is a vertically polarized antenna that, when used with VHF FM tactical radios, extends the range of transmission considerably and provides some degree of EP. The half rhombic antenna, when properly employed, decreases VHF FM radio susceptibility to hostile EW operations, and enhances the communications ranges of the deployed radio sets. This effect is realized by directing the maximum signal strength in the direction of the desired friendly unit.

9-111. The VHF half rhombic antenna is a high gain, lightweight, directional antenna. It operates over the frequency range of 30–88 MHz. The antenna and all the ancillary equipment (guys, stakes, tools, and mast sections) can be packaged in a carrying bag for manpack or vehicular transportation.

9-112. Figure 9-24 is an example of the OE-303 VHF half rhombic antenna. The planning range for the OE-303 is equivalent to the planning range of the OE-254. The OE-303 half rhombic antenna is used with the AB-1244 mast assembly, consisting of 12 tubular mast sections (five lower-mast sections, one mast transition adapter, five upper-mast sections, and antenna adapter), a mast base assembly, and assorted ancillary equipment. When erected, the mast assembly is stabilized by a two-level, four-way guying system.

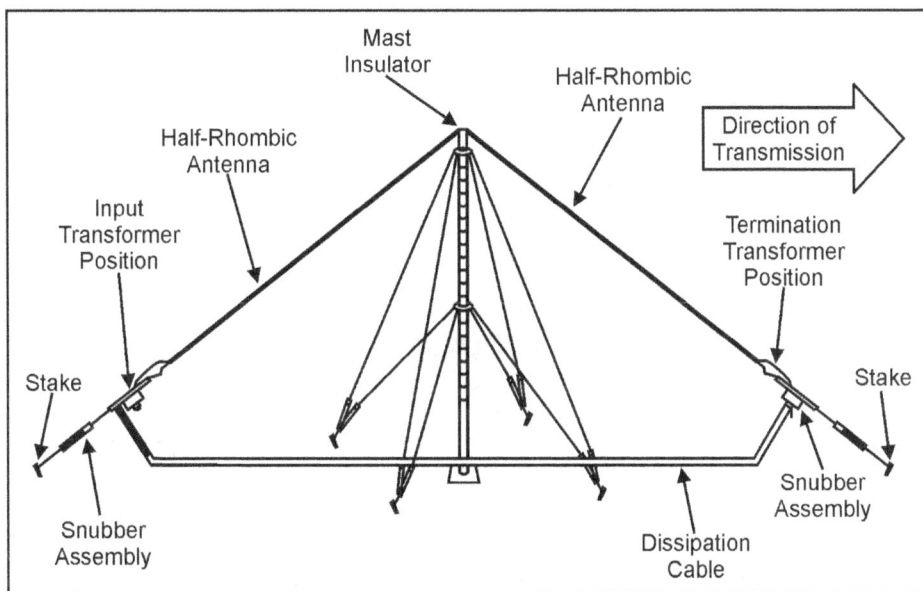

Figure 9-24. OE-303 half rhombic VHF antenna

9-113. The OE-303 antenna handles RF power levels up to 200 watts. It matches a nominal 50 ohm impedance with a VSWR of no more than 2:1, over the entire frequency range of the antenna. It meets the operation, storage, and transit requirements as specified in AR 70-38.

9-114. The OE-303 half rhombic antenna has the following characteristics and capabilities—

- Erected in a geographical area of 53.3 meters (175 ft) in diameter, or less, depending upon the frequency.
- Mounted on any structure approximately 15.2 meters (50 ft) in height.
- Azimuthal directional change within 1 minute.
- Transported by manpack or tactical vehicle when fitted into a package.
- Operation with the four-port FHMUX.

9-115. The OE-303 half rhombic antenna is used for special applications; it is task assigned as required. Its primary use is on C2 and intelligence nets to a higher headquarters. It must be available for use by units that habitually operate over extended distances from parent units, and must be available to units for special tasks. For more information on the half rhombic OE-303 antenna, refer to TM 11-5985-370-12.

High Frequency Antennas Usable at VHF and UHF

9-116.   Simple vertical half-wave dipole/doublet and quarter wave monopole antenna are very popular for omnidirectional transmission and reception over short range distances. For longer distances, rhombic antennas made of wire and somewhat similar in design to HF versions may be used to good advantage at frequencies as high as 1 GHz.

Dipole (Doublet) Antenna

9-117.   The dipole (doublet) is a half-wave antenna consisting of two quarter wavelength sections on each side of the center. It is also considered a center fed antenna. Figure 9-25 is an example of an improvised dipole (doublet) antenna used with FM radios.

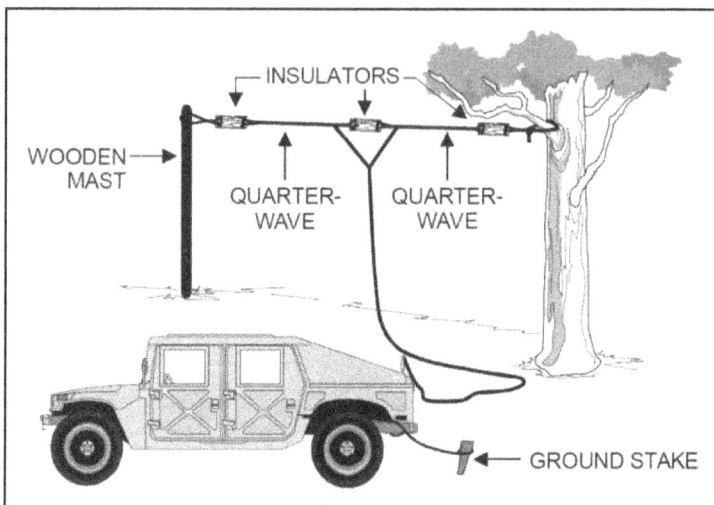

Figure 9-25. Half-wave dipole (doublet) antenna

9-118.   A transmission line is used for conducting electrical energy from one point to another, and for transferring the output of a transmitter to an antenna. Although it is possible to connect an antenna directly to a transmitter, the antenna generally is located some distance away. In a vehicular installation, for example, the antenna is mounted outside, and the transmitter inside the vehicle.

9-119.   Center-fed half-wave FM antennas can be supported entirely by pieces of wood. Figure 9-26 is an example of a horizontal (A) and vertical (B) center-fed half-wave antenna. These antennas can be rotated to any position to obtain the best performance. If the antenna is erected vertically, the transmission line should be brought out horizontally from the antenna, for a distance equal to at least one-half of the antenna's length, before it is dropped down to the radio set.

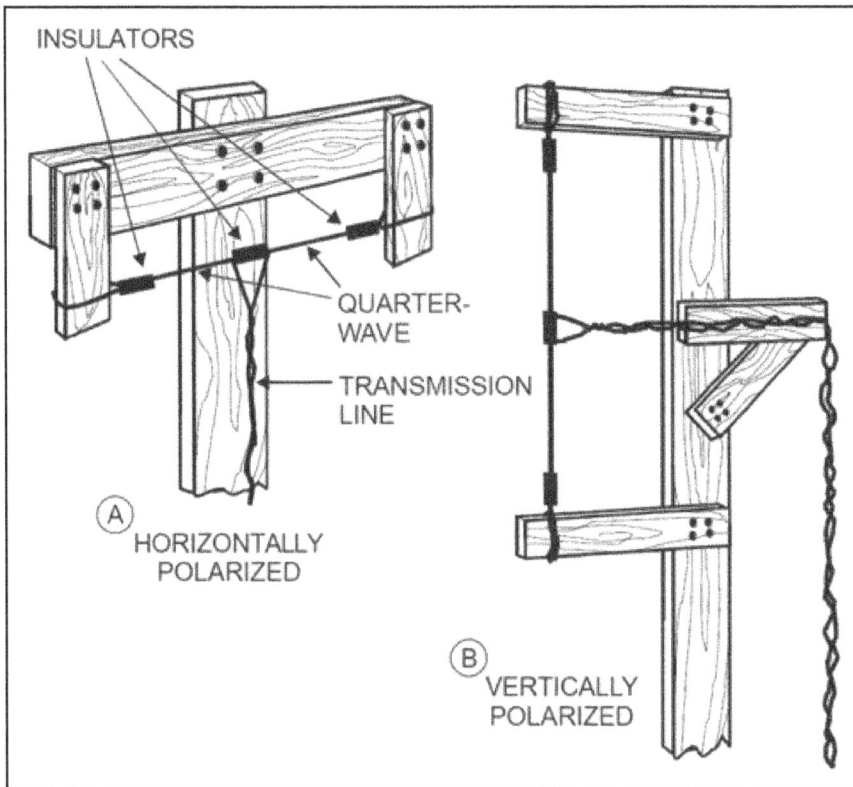

Figure 9-26. Center-fed half-wave antenna

9-120. Figure 9-27 is an example of an improvised vertical half-wave antenna. This technique is used primarily with FM radios. It is effective in heavily wooded areas to increase the range of portable radios. The top guy wire can be connected to a limb, or passed over the limb and connected to the tree trunk or a stake.

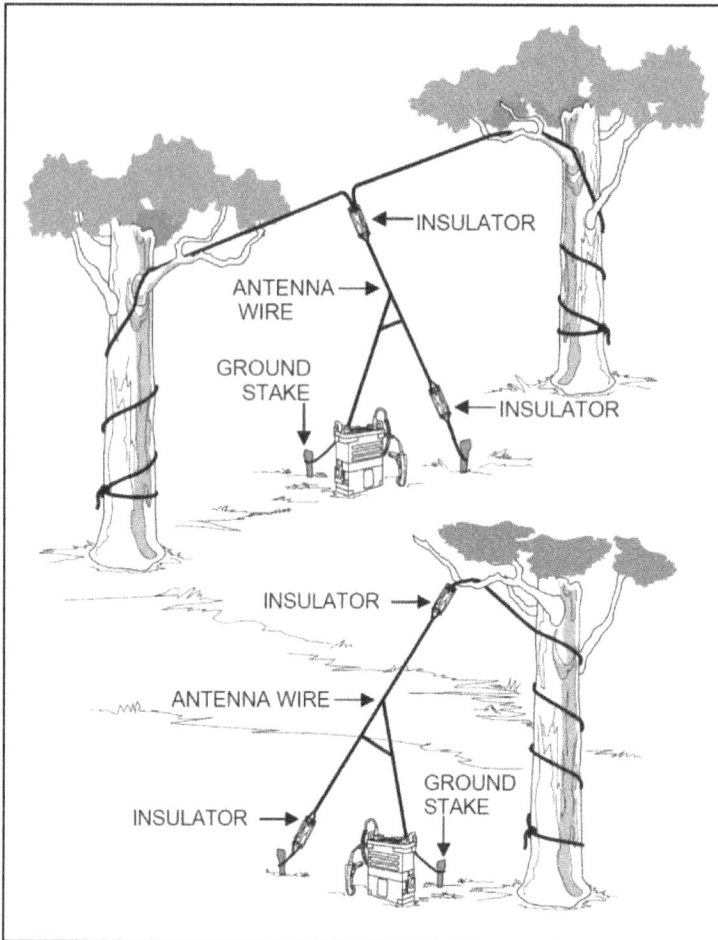

Figure 9-27. Improvised vertical half-wave antenna

SATELLITE COMMUNICATIONS ANTENNAS

9-121.   The most important consideration in siting LOS equipment is the antenna elevation with respect to the path terrain. Choose sites that exploit natural elevations.

Antenna Siting Considerations

9-122.   The most important consideration in siting over-the-horizon systems is the antenna horizon (screening angles) at the terminals. As the horizon angle increases, the transmission loss increases, resulting in a weaker signal.

9-123.   The effect of the horizon on transmission loss is very significant. Except where the consideration of one or more other factors outweighs the effect of horizon angles, the site with the most negative angle should be fist choice. If no sites with negative angles exist, the site with the smallest positive angle should be the first choice.

9-124.   The horizon angle can be determined by using a transit at each site and sighting along the circuit path. The on-site survey will determine the visual horizon angle. The radio horizon angle is slightly different from the visual horizon angle: however, the difference is generally insignificant.

9-125.   The horizon angle is measured between the tangent at the exact location of the antenna and a direct LOS to the horizon. The tangent line is a right angle (90 degrees) to a plumb line at the antenna site. If the LOS to the horizon is below the tangent line, the horizon angel is negative.

9-126.   Trees, building, hills or the Earth can block a portion of the UHF signals, causing an obstruction loss. To avoid signal loss due to obstruction and shielding, clearance is required between the direct LOS and the terrain. Path profile plots are used to determine if there is adequate clearance in LOS systems.

9-127.   Weak or distorted signals may result if the SATCOM set is operated near steel bridges, water towers, power lines, or power units. The presence of congested air-traffic conditions on the proximity of microwave equipment can result in significant signal fading, particularly when a non-diversity mode is employed.

9-128.   For LOS and TACSAT communications the AN/PSC-5 family of radios are the most widely used radios. The AN/PSC-5 provides LOS communications with the AS-3566 antenna and long range SATCOM with the AS-3567 and AS-3568 antennas. The following paragraphs describe several antennas and their characteristics.

AS-3566, Low Gain Antenna

9-129.   The AS-3566 has the following characteristics—
- **Frequency range (LOS):** 30–400 MHz.
- **DAMA:** 225–400 MHz.
- **Non DAMA:** 225–400 MHz.
- **Polarization:** directional.
- **Power capability:** determined by terminating resistor.
- **Azimuthal (bearing):** directional.

AS-3567, Medium Gain Antenna

9-130.   The AS-3567 (refer to Figure 9-28) has the following characteristics—
- **Frequency range:** 225–399.995 MHz.
- **Beam width:** 85 degrees.
- **Orientation:**
  - Directional.
  - Elevation (0–90 degrees).
- **Input impedance:** 50 ohms.
- **VSWR:** 1.5:1
- **Gain:**
  - 6 dB (225–318 MHz).
  - 5 dB (318–399.995 MHz).

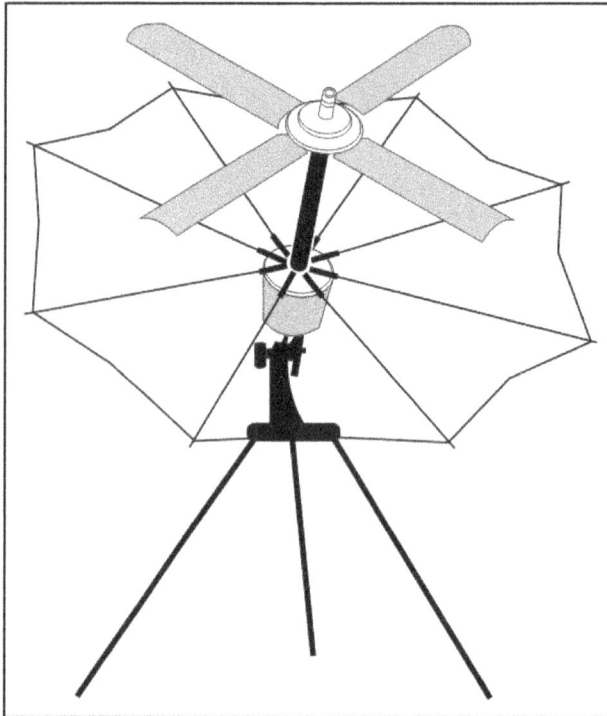

Figure 9-28. AS-3567, medium gain antenna

AS-3568, High Gain Antenna

9-131. The AS-3566 (refer to Figure 9-29) has the following characteristics—

- **Frequency range:** 240–400 MHz.
- **Beam width:** 77 degrees.
- **Orientation:**
  - Directional.
  - Elevation (0 to 90 degrees).
  - Azimuth+180 degrees.
- **Input impedance:** 50 ohms.
- **VSWR:** 1.5:1
- **Gain:**
  - 8 dB (240–318 MHz).
  - 6 dB (318–400 MHz).
- **Power:** up to 140 watts.

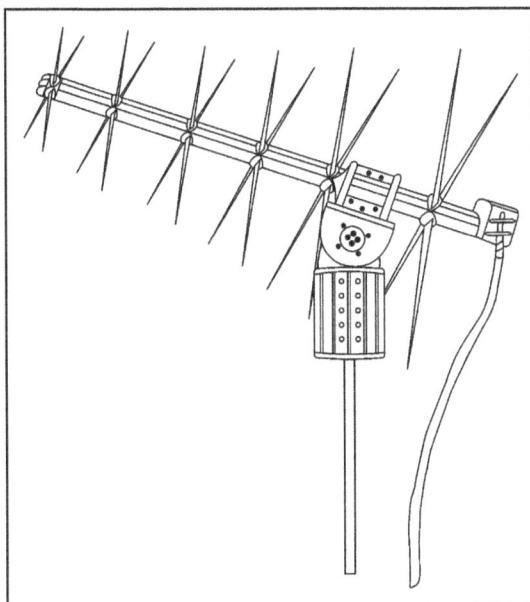

Figure 9-29. AS-3568, high-gain antenna

FIELD REPAIR

9-132.   Antennas that are broken or damaged cause poor communications or even communications failure. If a spare antenna is available, replace the damaged antenna. When a spare is not available, the user may have to construct an emergency antenna. The following paragraphs provide suggestions on repairing antennas and antenna supports.

REPAIR OF A WHIP ANTENNA

9-133.   A broken whip antenna can be temporarily repaired. If the whip is broken in two sections, rejoin the sections. Remove the paint and clean the sections this will help to ensure a good electrical connection. Place the sections together, secure them with a pole or branch, and lash them with bare wire or tape above and below the break (refer to Figure 9-30, antenna A).

9-134.   If the whip is badly damaged, use a length of field wire (WD-1/TT) the same length as the original antenna. Remove the insulation from the lower end of the field wire antenna, twist the conductors together, insert them in the antenna base connector, and secure with a wooden block. Use either a pole or a tree to support the antenna wire (refer to Figure 9-30, antenna B).

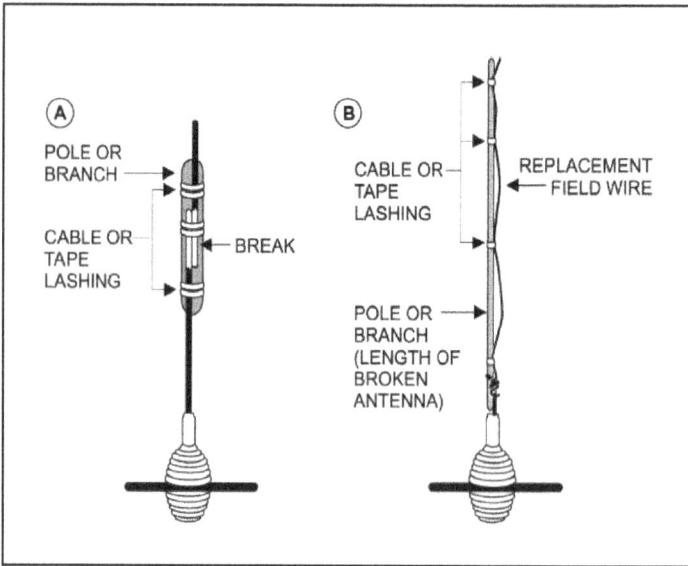

Figure 9-30. Field repair of broken whip antennas

WIRE ANTENNAS

9-135. Emergency repair of a wire antenna may involve the repair or replacement of the wire used as the antenna or transmission line. It may also involve the repair or replacement of the assembly used to support the antenna. When one or more antenna wires are broken, reconnecting the broken wires can repair the antenna. To do this, lower the antenna to the ground, clean the ends of the wires, and twist the wires together. When possible, solder the connection and reassemble.

9-136. Antenna supports may also require repair or replacement. A substitute item may be used in place of a damaged support and, if properly insulated, may consist of any material of adequate strength. If the radiating element is not properly insulated, field antennas may be shorted to ground, and be ineffective.

9-137. Many common items can be used as field expedient insulators. Plastic or glass (to include plastic spoons, buttons, bottlenecks, and plastic bags) is the best insulator. Wood and rope also act as insulators although they are less effective than plastic and glass (refer to Figure 9-31 for examples of field expedient antenna insulators). The radiating element, the actual antenna wire, should touch only the antenna terminal, and should be physically separated from all other objects other than the supporting insulator.

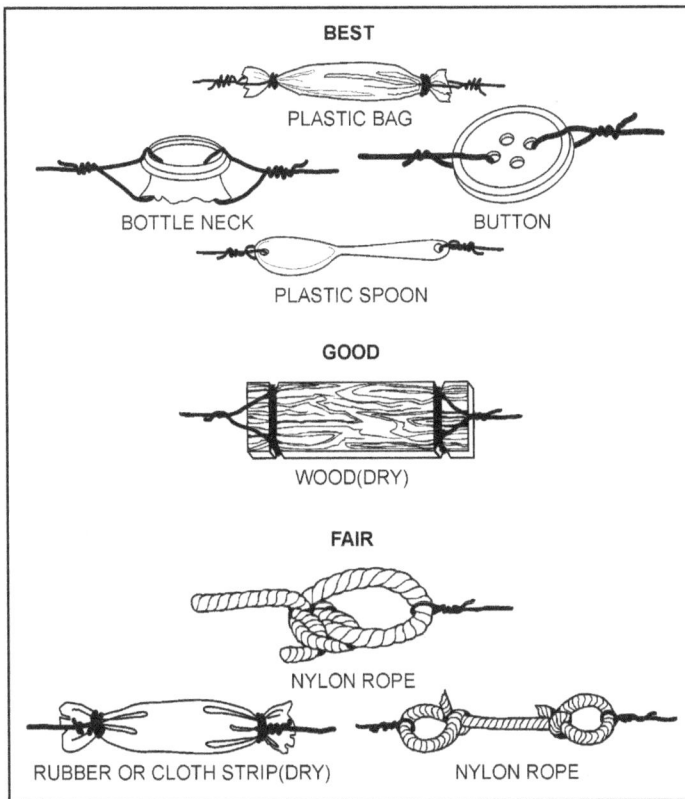

Figure 9-31. Examples of field expedient antenna insulators

ANTENNA GUYS

9-138. Guys stabilize the supports for an antenna. They are usually made of wire, manila rope, or nylon rope. Broken rope can be repaired by tying the two broken ends together. If the rope is too short after the tie is made, add another piece of rope or a piece of dry wood or cloth to lengthen it. Broken guy wire can be replaced with another piece of wire. To ensure that the guys made of wire do not affect the operation of the antenna, cut the wire into several short lengths and connect the pieces with insulators. Figure 9-32 shows an example of repaired guy lines with wood.

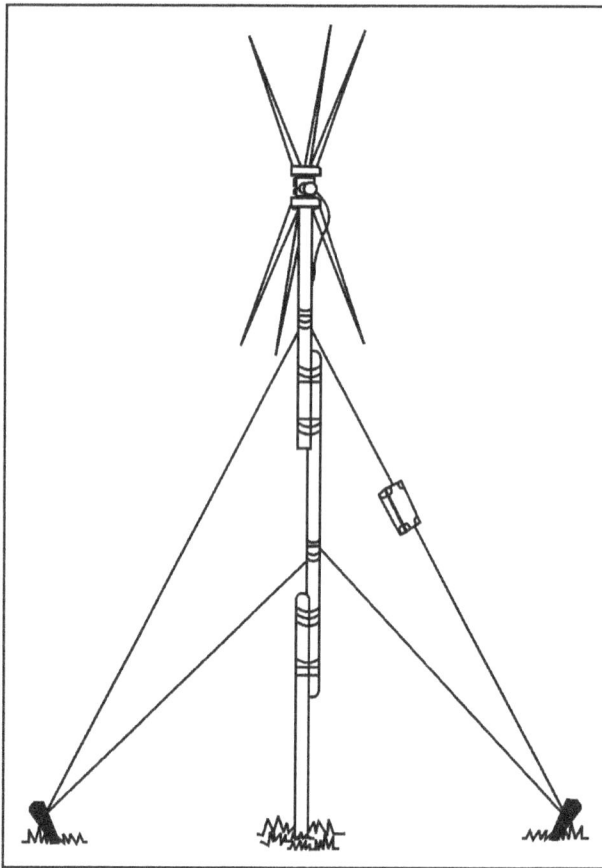

Figure 9-32. Repaired antenna guy lines and masts

Antenna Masts

9-139.   Masts support some antennas and if broken, one can be replaced with another of the same length. When long poles are not available as replacements, short poles may be overlapped and lashed together with rope or wire to provide a pole of the required length.

Chapter 10

Automated Communications Security Management and Engineering System

This chapter addresses the Automated Communications Security Management and Engineering System (ACMES) and its hardware and software components, designed to meet critical requirements to both decentralize and automate the process of generating and distributing data vital to communications systems. The ACMES supports the current version of the AKMS.

SYSTEM DESCRIPTION

10-1. The AKMS integrates all functions of cryptographic management and engineering, SOI, EP, and cryptographic key generation, distribution, accounting, and audit trail recordkeeping into a total system designated as the ACMES.

10-2. The ACMES provides commanders the necessary tools to work with the widely proliferating COMSEC systems associated with the MSE, JTIDS, EPLRS, SINCGARS, and other keying methods (electronic key generation, OTAR transfer, and electronic bulk encryption and transfer) being fielded by the Army.

10-3. The ACMES is a hardware and software system that provides the communications planner with the capability to design, develop, generate, distribute, and manage both decentralized and automated communications-electronics operating instructions (CEOI)/SOIs. ACMES can produce the EP fill variables to support SINCGARS in data file and electronic formats; it also produces SOI outputs in either electronic or hard copy (paper) formats. The objective is to fully utilize the electronic data storage devices (ANCD and SKL) to eliminate the need for exclusive use of a hard copy paper SOI.

10-4. The planning and distribution of ACMES products are essential to the success of military operations, and are a command responsibility. The controlling authority is the commander, who establishes a cryptographic net. Within divisions, brigades, and battalions, commanders may be assigned responsibilities depending upon command policy and operational situations. Table 10-1 outlines the ACMES functions and products at various command levels, theater to battalion.

10-5. Signal officers at corps and division (G-6s) levels (and separate brigades) use their ACMES components to design, develop, generate, and distribute CEOI and SINCGARS FH data, along with HF, UHF, and VHF frequency assignments at their respective levels and subordinate levels, as appropriate.

10-6. Brigades and separate battalion units use their ACMES components to selectively distribute generated CEOI and SINCGARS FH data for use at their respective and subordinate levels.

Note. Refer to AR 380-40, AR 380-5, AR 25-2, AR 380-53, and FM 6-02.72 for additional information on controlling authority and commanders' responsibilities regarding cryptographic networks.

Table 10-1. ACMES functions at various command levels

Command Levels	Media	Function
Theater	Disk	Generates pairs of operational TRANSEC keys every 30 days, for ICOM and non-ICOM SINCGARS. The communications systems directorate of a joint staff (J-6) generates the TRANSEC keys every 90 days.
Corps	Disk	Generates the sign/countersign, smoke/pyrotechnic signals, suffix/expander, hopsets, and CEOI/SOI at corps level; receives TRANSEC keys from theater.
Division	Disk/ANCD/SKL	Uses corps data, or if authorized, generates and merges SOI data, generates COMSEC data (division TEK), and generates FH data (NET IDs and division TSKs).
Brigade	Disk/ANCD/SKL	Receives the generated CEOI/SOI and other data, such as hopsets and TRANSEC keys from division.
Battalion	Disk/ANCD/SKL/ Paper and ECCM Fill Device	Receives the CEOI/SOI information and other data such as hopsets and TRANSEC keys from the brigade.

Note. In some situations, theater may not be the highest level of command to generate TRANSEC keys. It depends on the mission, situation and if the unit is a supporting command.

HARDWARE

10-7. The following paragraphs address ACMES hardware components (AN/GYK-33A, lightweight computer unit [LCU], LCU printer, random data generator [RDG], and ANCD). These hardware components, along with the software, make up the ACMES workstation. Workstations with the RDG are organic to corps, divisions, and separate brigades. (Refer to FM 6.02-72 for more information on the ACMES workstation.)

AN/GYK-33A, LIGHTWEIGHT COMPUTER UNIT

10-8. The LCU is a computer system that may serve as a host for many application software packages (programs) designed to provide the user with the means to accomplish assigned missions. Figure 10-1 is an example of a LCU.

10-9. When operating the ACES and ACES DTD application software, the LCU provides the user with the capability to generate, store, print, and/or electronically transfer both SC and FH information. It also provides the TSK for EP. These capabilities are designed to be more responsive to rapidly changing and highly mobile conditions on the battlefield.

10-10. The LCU consists of a computer with a keyboard and a crystal display. The 10-inch display normally shows 25 80 character lines of alphanumeric information. The video graphics array display contains 640 x 480 pixels, and supports 16 levels of shading. The LCU may be used at a fixed workstation or may be carried to most locations when using battery power. The LCU holds 20 rechargeable nickel cadmium or alkaline batteries (size C). Mission duration under battery power is less than two hours, at 70 degrees ambient temperature, unless batteries are recharged or replaced. A typical workstation setup might require space for peripheral devices such as a printer, printer paper, interface transfer cables, and/or interfaced devices (for example, SINCGARS and ANCDs).

Figure 10-1. Lightweight computer unit

LIGHTWEIGHT COMPUTER UNIT PRINTER

10-11. The LCU printer is a small, lightweight dot-matrix printer that is easily transportable. The LCU printer has a print rate of 160 characters per second in the draft printing mode, and 80 characters per second in the near-letter-quality printing mode. It is powered by either battery, or the LCU power supply. The printer ribbon is capable of printing several hundred pages; it is disposable, and easily replaced. The normal line width is 80 characters. The printer will accept paper widths from three to 8.5 inches, and has a tractor feed attachment that accepts 8.5 x 11 inch continuous form, fan-fold paper. The printer operates on 9–36 VDC (battery and vehicular), or 110 VAC.

RANDOM DATA GENERATOR, AN/CSZ-9

10-12. The RDG provides the LCU with the necessary random data to allow the ACES software to generate SOI and/or TSK fill data. Figure 10-2 shows the RDG. It is a controlled cryptographic item, and must be transported as authorized by AR 380-5.

Note. RDGs are only authorized and issued for use at selected echelons.

10-13. The RDG is a self-contained unit, powered by five D-size 1.5 volt batteries, located beneath the bottom shelf/foot plate of the unit. Serviceable batteries should be installed prior to using the unit.

10-14. The on/off switch on the front panel activates the RDG; however, the unit does have a sleep mode that inactivates the unit when not in use for an extended period, to conserve energy drain from the unit's battery power supply. The unit is provided with a cable that connects the unit (from its rear panel) to a serial port on the computer.

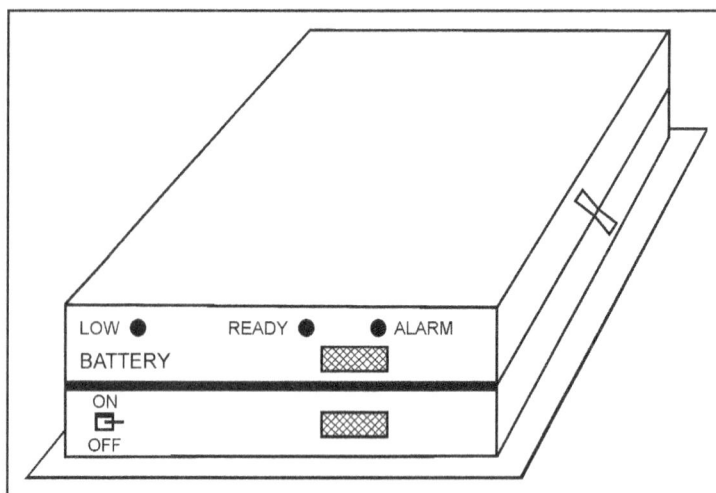

Figure 10-2. Random data generator

SOFTWARE

10-15.   Revised DTD software (RDS) and the ACES application software make up the applications software for ACMES.

10-16.   The RDS software is unclassified, pre-installed on the ANCD, and designated to provide support of user's needs with regard to SOIs and FH data for the FH SINCGARS. RDS is divided into a CEOI/SOI portion and a SINCGARS portion.

10-17.   The CEOI/SOI portion provides the capability to receive, store, display, and transfer CEOI/SOI data.

10-18.   The SINCGARS portion of RDS provides the capability to fill SINCGARS, and receive, store, display, and transfer the data that is required to fill these radios.

AUTOMATED COMMUNICATIONS ENGINEERING SOFTWARE

10-19.   ACES is a net planning software program that replaced the Revised Battlefield Electronic Communications-Electronics Operational Instruction/Signal Operating Instructions System (RBECS), for the US Army. ACES works in a ruggedized Windows NT COTS platform for tactical operations as well as in desktop Windows NT workstations in strategic locations. ACES allows military users to perform fully automated cryptographic net, SOI, CEOI, joint CEOI and EP planning, management, validation and generation distribution at the time and location needed.

10-20.   The network planning functionality of ACES incorporates cryptonet planning, key management, and key tag generation. The planning concept relates to the development of network structures supporting missions and plans. The data for a given plan includes individual nets, which are assigned individual net members. Net members are associated with a specific platform and equipment. Once net members, platforms, and equipment are designated, specific equipment fill locations are defined and key tags/keys are associated with the equipment locations.

10-21.   The equipment records, which include platform data, net data, and key tags, are then downloaded to the DTD, and subsequently associated with the required key. Similarly, the EP data and SOI are generated by the ACES workstation operator and can be downloaded to the DTD.

MASTER NET LIST

10-22.   Master net list (MNL) maintains all nets requiring SOI assignments. Maintaining the MNL is essential to creating deconflicted SOI assignments. Additionally, nets that have been created or imported may be edited from the MNL allowing individual frequency assignments to be tracked with assigned equipment. The ACES version of the MNL has direct correlation to standard frequency action format (SFAF) line item numbers, so as you create the MNL the base for the SFAF and the SOI is being compiled at the same time.

10-23.   The MNL is the database link for all information listed under a plan, such as nets, frequencies and equipment. The MNL provides the capability to create, edit, organize, and delete nets. Before creating the MNL, the ACES workstation operator must know how many nets are required, what types of equipment will be used, and specific information about the equipment, such as maximum transmit power, frequency bands, and emission designators. This information is available from the spectrum or area frequency manager. This section provides the information in creating the MNL folder and entering and managing the information within the folder. (Refer to TB 11-7010-293-10-2 for more detailed information on ACES and how to build a MNL.)

10-24.   The MNL module of the ACES software may also be displayed in service specific views (US Army, USN, USAF, and USMC) or joint combined. The MNL also incorporates a number of SFAF compatible fields to facilitate the transfer of data to and from other frequency management systems such as Spectrum XXI, as well as service unique systems. The database capabilities of the ACES workstation allow the data in the MNL to be used to create the initial SFAF frequency proposal and the SOI.

10-25.   The ACES software components on the ACES workstation include the ACES core module, general purpose module, resource manager module, MNL module, SOI module, and CNR module.

COMBAT NET RADIO MODULE

10-26.   The CNR module provides the necessary functions and procedures to create and modify hopsets, loadsets, and to generate SINCGARS TSKs. It also provides the capability to plan CNR nets in all bands. CNR net planning is integrated with the MNL module.

RESOURCE MANAGER MODULE

10-27.   The resource manager module contains frequency resources and allows these resources to be created, edited, merged, deleted, and printed. The resource manager also provides the planner the capability to import and export resources in RBECS, ACES, integrated system control, and SFAF formats.

SIGNAL OPERATING INSTRUCTIONS MODULE

10-28.   The SOI module allows editions to be created and updated. Each SOI is identified by a short title and edition and may contain up to ten time periods. SOI is a series of orders issued to control and coordination of the signal operations of a command or activity. It provides guidance needed to ensure the speed, simplicity, and security of communications. Nets are selected from the MNL to be included in a generated SOI edition. Before the SOI can be generated the MNL must be saved and validated. Figure10-3 is an example of an expanded ACES navigation tree and Figure 10-4 is an example of the general sequence for planning a CNR net.

GP-General Purpose
Allows the cryptonet
plans of Nets. Platforms
and Equipment not
defined in ACES

Area Common User System
Plan Cryptonet Network

ARC-220 Folder
Create ALE
Addresses/Equipment

Master Net List (MNL)
Maintains all nets
Base for SFAF Proposal and
SOI Create SFAF and SOI
Generation commands are
accessed through the MNL.

Resource Manage
Manages all frequencies

SATCOM Folder
Support Crypto Planning
for SCAMP and SMART-T

SOI Folder
Contains Dictionaries
Groups, Short Titles and
Editions. Additional SOI
data is built under this
component.

Combat Net Radio (CNR)
Generate Hopsets
Manually create Loadsets
Create TEK/Share Groups

LEGEND

ACES	Automated Communications Engineering Software	SFAF	Standard Frequency Action Format
ALE	Automatic Link Establishment	SOI	Signal Operating Instructions
SATCOM	Satellite Communications	TEK	Traffic Encryption Key
SCAMP	Single-channel Anti-jam Man Portable		

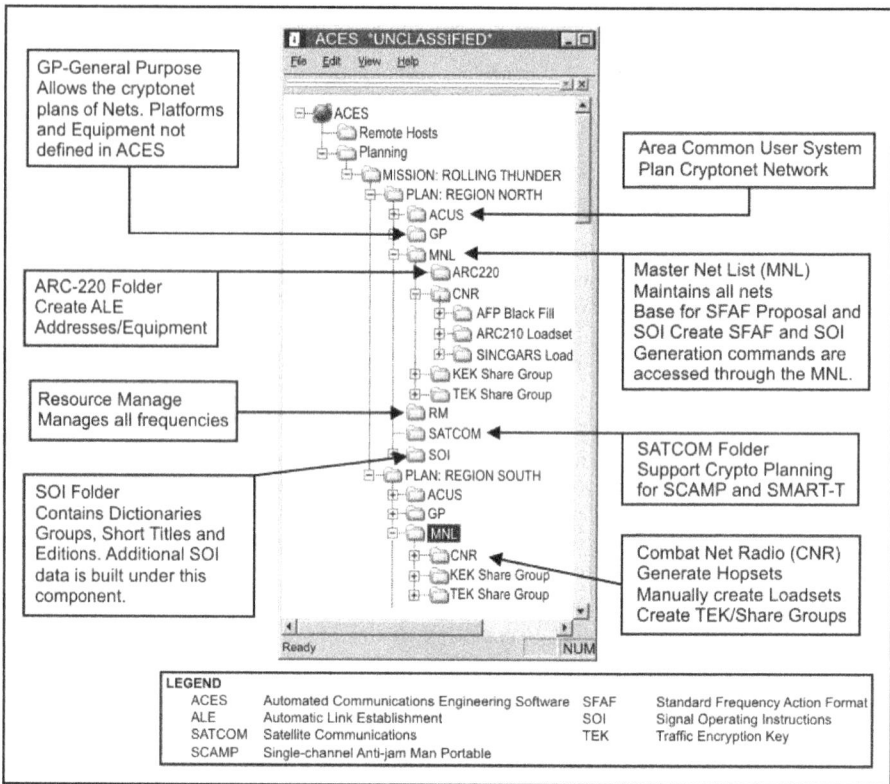

Figure 10-3. Expanded ACES navigation tree

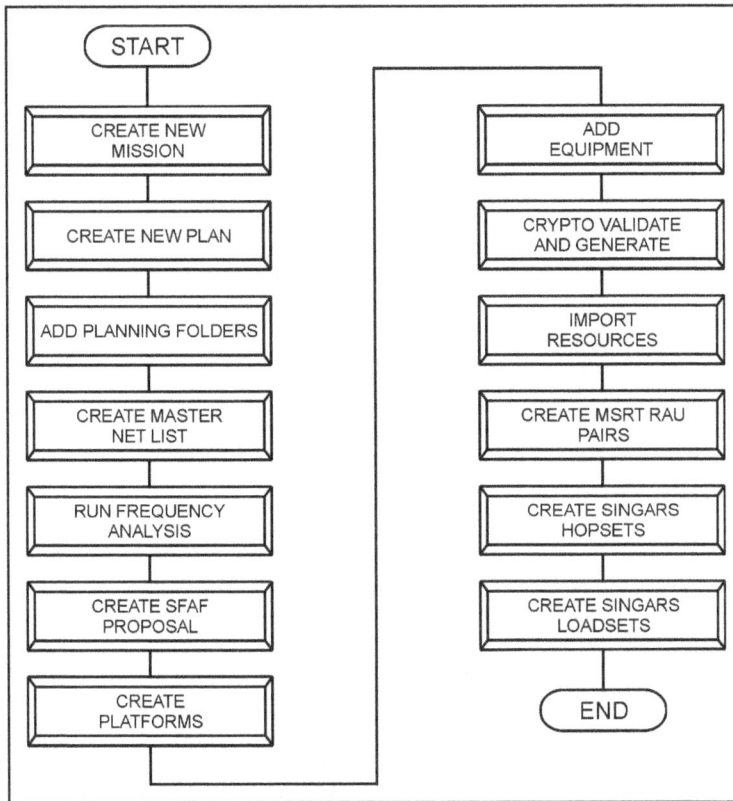

Figure 10-4. Example for planning a CNR net

COMMUNICATIONS-ELECTRONICS OPERATING INSTRUCTIONS/SIGNAL OPERATING INSTRUCTIONS DEVELOPMENT

10-29.   ACES is designed to decentralize and automate CEOI/SOI generation. Generating and distributing ACES CEOI/SOI can be done with virtually no dependence on the NSA. ACES is also capable of building a division size CEOI/SOI in two to five hours. The NSA normally requires 60–90 days lead time; a manual build normally requires three to five days to produce the same CEOI/SOI. ACES can respond quickly to a compromise of CEOI/SOI in the field, or to rapidly changing force structures and can regenerate frequencies and net call signs in three to five hours (depending on database size).

10-30.   Although ACES automates the generation process, the signal officer must first design the CEOI/SOI on paper. Table 10-2 lists the initial steps for designing and developing CEOI/SOI data. The following paragraphs provide more detail on CEOI/SOI development.

Table 10-2. Initializing ACES CEOI/SOI data

Step	Description
1	Research and extract data from the modified table of organization and equipment, which authorizes the use of personnel and equipment.
2	Determine the doctrine to be followed.
3	OPORD/OPLAN/unit SOP.
4	Frequency list from the spectrum manager.
5	Determine how many nets and frequencies are required. Use the current CEOI/SOI as a starting point.

FREQUENCY ASSIGNMENT

10-31. ACES can be used for frequency and net call sign generation in all frequency bands currently used by the military. All frequency assignments are based on the authorized frequencies of the using organization. The available frequencies are listed in the current resource frequency allocation. The initial step in preparing the net/frequency assignment plan is to identify the unit nets required for C2 of tactical operations.

10-32. After all nets are identified, compare the resulting frequency requirements with the number of frequencies available. Spare frequencies are available for assignment in most CEOI/SOI, with their use being controlled by the major organization's (controlling authority) signal officer. If the frequency allocations and assignments are inadequate, additional frequencies must be requested through a higher command or area frequency coordinator, or some nets will be required to share frequencies with other nets.

10-33. Various types of frequency assignments should be considered when developing the database to generate a CEOI/SOI. Ideally, frequencies are randomly assigned to nets, designed to receive a changing frequency with each change in time period. A net may be sufficiently important to warrant a dedicated (sole user) frequency for its use. The frequency is unique to the organization, and is only used by one net during any time period. This assignment is reserved for C2 nets. (Refer to Table B-1 for an outline of the frequency bands.)

Fixed Frequencies

10-34. Fixed frequencies are usually assigned to nontactical units. The frequency value is manually assigned, and is unique to the specific net. The frequency value is non-changing for all time periods of the generation. For example, the medical evacuation net may be assigned 34.000 MHz, and this frequency will never change; also, 34.000 MHz will not be assigned to another net. Fixed frequencies do not have any restrictions assigned, and are used on SC nets only.

Reusing and Sharing Frequencies and Frequency Separation

10-35. The lack of available frequencies, or abundance of needed nets, may require nets to reuse or share frequencies. Also, some nets require frequency separation from other nets, to prevent interference.

10-36. When the number of nets requiring frequencies is greater than the number of available frequencies, frequency reuse (common user) may be necessary. Nets are selected for inclusion in a reuse plan on the basis of low operating power, geographic separation, terrain masking, and other factors permitting the use of the same frequencies on a noninterference basis.

10-37. The following list of nets should be excluded from a reuse plan—

- Command and wireless network extension (and corresponding) nets.
- Command and fire control nets (maneuver units).
- Fire direction nets (division artillery).
- Any FM aviation net (cavalry/attack).
- Any emergency net.
- Any spare net.
- Any anti-jam or alternate net.

10-38. Sharing frequencies is another method of reducing the number of frequencies required; two or more nets use a shared frequency. The sharing nets will receive the same frequency for a given time period, either fixed or discrete. Typical nets that use shared frequencies are survey and weather nets.

10-39. A separation plan provides a frequency separation between nets. This plan is used when operating more than one net within a communications van, in close proximity to other nets, or when mutual frequency interference may result between radios. A separation plan is designed to allow these nets to communicate simultaneously without mutual interference. Frequency assignment separation requirements include co-site, wireless network extension, and other alternate nets.

10-40. Mutual interference problems may result if FM transmitters operating on different frequencies are situated in the same locale. To effectively reduce these interference problems, ACES adheres to the following basic standards—

- Frequencies with an exact separation factor of 5.750 or 23.000 MHz to collocated nets are not assigned.
- Frequencies that are on the order of the second harmonic. (For example, the frequency setting of 30.000, 32.650, and 35.000 MHz will possibly interfere with radios using 60.000, 65.300, and 70.000 MHz, respectively, are not assigned.)

SINCGARS Spectrum Management Variables

10-41. The G-6/S-6 section identifies requirements for the construction of loadsets to support the radios that are employed by their organization. These loadsets, once defined, are then constructed using ACES, saved to file, and distributed to subordinate organizational units or elements for follow-on distribution to respective users. The construction of loadsets is defined by the user, and is primarily based upon the identification of the nets that the radio user is required to enter/monitor.

10-42. For example, the commander of an infantry battalion would normally be a member of several FH SINCGARS nets. One of the commander's SINCGARS could require the following to be loaded—

- Brigade command net.
- Brigade operations net.
- Battalion command net.
- Battalion operations net.
- Brigade wireless network extension net.

10-43. RTOs will normally load all six preset channels on the SINCGARS, with operational NET IDs and TEKS. If a requirement to perform an OTAR arises, all stations involved with OTAR must load a KEK (stored in the ANCD) into preset Channel 6 on the SINCGARS, with an appropriate NET ID.

Loadset Updates

10-44. The responsible signal section personnel using ACES and RDS, as appropriate, maintain loadset data. Loadset data is updated with new replacement key data, when appropriate, before the current key expires. The loadset data is then saved to file, and distributed to users via ANCD/SKL, so they are in place and available for loading into the SINCGARS at the appropriate key changeover time. Additionally, the signal sections should have several sets of loadsets with associated keys, already constructed and distributed (or available for expeditious distribution) for immediate use.

Loadset Revisions/Creations

10-45. Existing loadsets may require revision when the required net content changes (unit reassignment or attachment). New loadsets may have to be constructed to meet new requirements (for example, a new task force organization is created).

Joint Automated CEOI System

10-46. The Military Communications Electronics Board has designated ACES as the Joint Spectrum Management Planning software. For multi-service operations it is called Joint Automated Communications-Electronics Operating Instructions System (JACS). JACS has the same basic function as ACES. JACS core purpose is to allow an interface between the joint CEOI generation tool with service unique communications planning software and spectrum management automated tools.

This page intentionally left blank.

Chapter 11

Communications Techniques: Electronic Protection

This chapter addresses EW and the EP techniques used to prevent enemy jamming and intrusion into friendly communications systems. It also addresses EP responsibilities, the planning process, signal security, emission control, preventive and remedial EP techniques and the Joint Spectrum Interference Resolution (JSIR) reporting procedures and requirements.

ELECTRONIC WARFARE

11-1. EW uses electromagnetic energy to determine, exploit, reduce, or prevent hostile use of the electromagnetic spectrum; it also involves actions taken to retain friendly use of the electromagnetic spectrum. Table 11-1 lists the three elements of EW.

Table 11-1. Electronic warfare elements

Element	Responsibilities
Electronic warfare support (ES)	Involves actions taken to search for, interrupt, locate, record, and analyze radio signals for using such signals in support of military operations.
	Provides EW information required to combat electronic countermeasures, to include threat detection, warning, avoidance, target location, and homing.
	Produces signals intelligence (SIGINT), communications intelligence, and electronic intelligence.
EA	Involves using electromagnetic or directed energy to attack personnel, facilities, or equipment with the intent of degrading.
	Includes actions taken to prevent or reduce the enemy's effective use of his frequencies; includes jamming and deception.
	Employs weapons that use either electromagnetic or directed energy as their primary destructive mechanism (lasers, RF weapons, and particle beams).
EP	Ensures friendly effective use of frequencies, despite the enemy's use of EW.
	Provides defensive measures used to protect friendly systems from enemy EW activities, such as— <ul><li>Careful siting of radio equipment.</li><li>Employment of directional antennas.</li><li>Operations using lowest power required.</li><li>Staying off the air unless absolutely necessary.</li><li>Using a random schedule, if one is used.</li><li>Using good radio techniques and continued operation.</li></ul> *Note.* Refer to Appendix H for more information on antenna placement and co-site interference.

ELECTRONIC WARFARE IN COMMAND AND CONTROL ATTACK

11-2. EW support, EA, and EP contribute to C2-attack operations. ES, in the form of combat information, can provide real-time information required to locate and identify adversary C2 nodes, and supporting/supported early warning and offensive systems during C2 attack missions. It produces SIGINT, and can provide timely intelligence about an adversary's C2 capabilities and limitations that can be used to update previously known information about the adversary's C2 systems. This updated information can be used to plan C2 attack operations, and provide damage assessment feedback on the effectiveness of the overall C2 warfare plan.

11-3. EA is present in most C2 attack operations in a combat environment. It includes jamming and electromagnetic deception or destruction of C2 nodes, with directed-energy weapons or anti-radiation missiles.

11-4. EP protects the electromagnetic spectrum for friendly forces. Coordinating the use of the electromagnetic spectrum through the joint restricted frequency list (JRFL) is a means of preventing fratricide among friendly electronic emissions. Equipment and procedures designed to prevent adversary disruption or exploitation of the electromagnetic spectrum are the best means friendly forces have to ensure their own uninterrupted use of the electromagnetic spectrum during C2 attack operations. (For more information on joint EW refer to JP 3-13.1.)

ELECTRONIC WARFARE IN COMMAND AND CONTROL PROTECT

11-5. The three elements of EW can also contribute to friendly C2 protect efforts. ES, supported by SIGINT data, can be used to monitor an impending adversary attack on friendly C2 nodes. In the form of signal security monitoring, ES can be used to identify potential sources of information for an adversary to obtain knowledge about friendly C2 systems.

11-6. EP can be used to defend a friendly force from adversary C2-attack. EP should be used in C2 protect to safeguard friendly forces from exploitation by adversary ES/SIGINT operations. Frequency management using the JRFL is essential to a successful coordinated defense against adversary C2-attack operations.

ADVERSARY COMMAND AND CONTROL ATTACK

11-7. Understanding the threat to the electromagnetic spectrum is the key to practicing sound EP techniques. Adversary C2 attack is the total integration of EW and physical destruction of resources, to deny friendly forces the use of electronic control systems. Potential adversaries consider C2 attack integral to all combat operations. They have invested in developing techniques and equipment to deny their enemies the effective use of the electromagnetic spectrum for communications.

11-8. Adversary C2 attack disrupts or destroys at least 60 percent of the command, control, intelligence, and weapons systems communications (30 percent by jamming and 30 percent by destructive fires). To accomplish this goal, enemy forces expend considerable resources gathering combat information about their enemies. As locations are determined, and units are identified, enemy forces establish priorities to—

- Jam communications assets.
- Deceptively enter radio nets.
- Interfere with the normal flow of their enemy's communications.

COMMANDERS ELECTRONIC PROTECTION RESPONSIBILITIES

11-9. EP is a command responsibility. The more emphasis the commander places on EP, the greater the benefits, in terms of casualty reduction and combat survivability, in a hostile environment. Because adversary C2 attack is a real threat on the modern battlefield, commanders at all levels must ensure their units are trained to practice sound EP techniques.

11-10. Commanders must constantly measure the effectiveness of the EP techniques; they must also consider EP while planning tactical operations. Commanders' EP responsibilities are—

- Review all after action reports where jamming or deception was encountered, and assess the effectiveness of defensive EP.
- Ensure all encounters of interference, deception, or jamming are reported and properly analyzed by the G-6/S-6 and the assistant chief of staff, intelligence (G-2) or intelligence staff officer (S-2).
- Analyze the impact of enemy efforts to disrupt or destroy friendly C2 communications systems on friendly OPLANs.
- Ensure the unit practices COMSEC techniques daily. Units should—
  - Change net call signs and frequencies often (in accordance with the SOI).
  - Use approved encryption systems, codes, and authentication systems.
  - Control emissions.
  - Make EP equipment requirements known through quick reaction capabilities that are designed to expedite procedure for solving, research, development, procurement, testing, evaluation, installations modification, and logistics problems as they pertain to EW.
  - Ensure radios with mechanical or electrical faults are repaired quickly; this is one way to reduce radio distinguishing characteristics.
  - Practice net discipline.

STAFF ELECTRONIC PROTECTION RESPONSIBILITIES

11-11. The staff is organized to assist the commander in accomplishing the mission. Specifically, the staff responds immediately to the commander and subordinate units. The staff should—

- Keep the commander informed.
- Reduce the time to control, integrate, and coordinate operations.
- Reduce the chance for error.

11-12. All staff officers provide information, furnish estimates, and provide recommendations to the commander; prepare plans and orders for military operations; and supervise subordinates to achieve mission accomplishment. Staff members should assist the commander in carrying out communications EP responsibilities. Specific responsibilities of the staff officers are—

- G-2/S-2—advises the commander of enemy capabilities that could be used to deny the unit effective use of the electromagnetic spectrum. They also keep the commander informed of the unit's signal security posture.
- G-3/S-3—exercises staff responsibility for EP and includes ES and EA scenarios in all CP and field training exercises, and evaluates EP techniques employed. They also include EP training in the unit training program.
- G-6/S-6—prepares and conducts the unit EP training program; ensures there are alternate means of communications for those systems most vulnerable to enemy jamming; ensures available COMSEC equipment is distributed to those systems most vulnerable to enemy information gathering activities and ensures measures are taken to protect critical friendly frequencies from intentional and unintentional interference. The G-6/S-6 also enforces proper use of radio, EP, and TRANSEC procedures on communications channels; performs frequency management duties, and issues SOIs on a timely basis; prepares and maintains a restricted frequency list of taboo, protected, and guarded frequencies and prepares the EP and restricted frequency list appendices to the signal annex with appropriate cross-references to the other annexes (EW, operations security, and deception) and to the SOI for related information.

PLANNING PROCESS

11-13. Threats to friendly communications must be assessed during the planning process. Planning counters the enemy's attempts to take advantage of the vulnerabilities of friendly communications systems.

As a minimum, four categories of EP planning must be considered: deployment, employment, replacement, and concealment. The following paragraphs address the deployment phase of the EP planning process.

GEOMETRY

11-14. Analyze the terrain, and determine methods to make the geometry of the operations work in the favor of friendly forces. Adhering rigidly to standard CP deployment makes it easier for the adversary to use the direction finder and aim his jamming equipment at his enemies.

11-15. Deploying units and communications systems perpendicular to the forward line of own troops (FLOT) enhance the enemy's ability to intercept communications because US forces aim transmissions in the enemy's direction. When possible, friendly forces must install terrestrial LOS communications parallel to the FLOT. This supports keeping the primary strength of US transmissions in friendly terrain. Refer to Figure 11-1 for an example of geometry during operations.

11-16. SC TACSAT systems reduce friendly CP vulnerability to enemy direction efforts. Tactical SATCOM systems are relieved of this constraint because of their inherent resistance to enemy direction finder efforts. Terrain features should be used when possible to mask friendly communications from enemy positions. This may mean moving senior headquarters farther forward and using more jump or TAC CPs so that commanders can continue to direct their units effectively.

11-17. Locations of CPs must be carefully planned, as CP locations generally determine antenna locations. The proper installation and positioning of antennas around CPs is critical. Antennas and emitters should be dispersed and positioned at the maximum remote distance, terrain dependent, from the CP, so that all of a unit's transmissions are not coming from one central location.

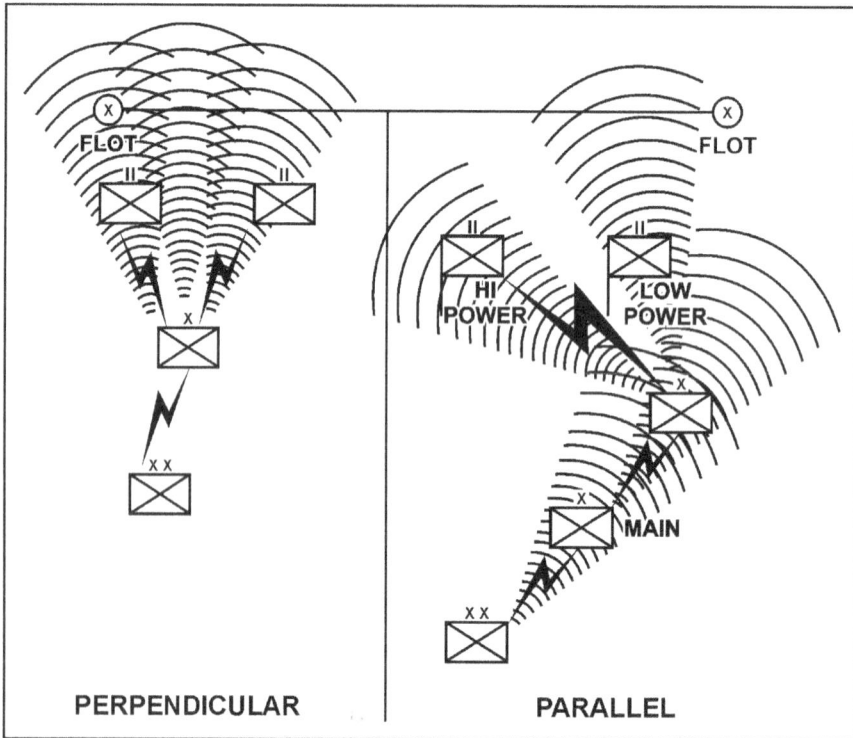

Figure 11-1. Geometry during operations

SYSTEM DESIGN

11-18.   Alternate routes of communications must be established when designing communications systems. This involves establishing sufficient communications paths to ensure that the loss of one or more routes will not seriously degrade the overall system. The commander establishes the priorities of critical communications links; the higher priority links should be afforded the greatest number of alternate routes.

11-19.   Alternate routes enable friendly units to continue to communicate despite the enemy's efforts to deny them the use of their communications systems. They can also be used to transmit false messages and orders on the route that is experiencing interference, while they transmit actual messages and orders through another route or means. A positive benefit of continuing to operate in a degraded system is the problematic degraded system will cause the enemy to waste assets that might otherwise be used to impair friendly communications elsewhere.

11-20.   Three routing concepts, or some permutation of them, can be used in communications—

- **Straight-line system**—provides no alternate routes of communications.
- **Circular system**—provides one alternate route of communications.
- **Grid system**—provides as many alternate routes of communications as can be practically planned.

11-21.   Avoid establishing a pattern of communications. Adversary intelligence analysts are highly trained to extract information from the pattern, and the text, of friendly transmissions. If easily identifiable patterns of friendly communications are established, the enemy can gain valuable information.

11-22.   The number of friendly transmissions tends to increase or decrease according to the type of tactical operation being executed. This deceptive communications traffic can be executed by using either false peaks, or traffic leveling. False peaks are used to prevent the enemy from connecting an increase of communications with a tactical operation. Transmission increases, on a random schedule, create false peaks.

11-23.   Tactically, traffic leveling is accomplished by designing messages to be sent when there is a decrease in transmission traffic. Thus, traffic leveling is used to keep the transmission traffic fairly constant. Messages transmitted for traffic leveling or false peaks must be coordinated to avoid operational security violations, mutual interference, and confusion among friendly equipment operators.

11-24.   ACES equipment, software, and subsequent SOI development resolves many problems concerning communications patterns; they allow users to change frequencies often, and at random. This has long been recognized as a key in confusing enemy traffic analysts. Adversary traffic analysts are confused when frequencies, net call signs, locations, and operators are often changed. The adversary uses US TTP to help perform their mission. Therefore, these procedures must be flexible enough to avoid establishing communications patterns.

REPLACEMENT

11-25.   Replacement involves establishing alternate routes and means of doing what the commander requires. FM voice communications are the most critical communications used by the commander during adversary engagements. As much as possible, critical systems should be reserved for critical operations. The adversary should not have access to information about friendly critical systems until the information is useless.

11-26.   Alternate means of communications should be used before enemy engagements. This ensures the adversary cannot establish a database to destroy primary means of communications. Primary systems must always be replaced with alternate means of communications, if the primary means become significantly degraded. These replacements must be preplanned and carefully coordinated; if not, the alternate means of communications could be compromised, and become as worthless as the primary means. Users of communications equipment must know how and when to use the primary and alternate means of communications. This planning and knowledge ensures the most efficient use of communications systems.

CONCEALMENT

11-27. OPLANs should include provisions to conceal communications personnel, equipment, and transmissions. It is difficult to effectively conceal most communications systems; however, installing antennas as low as possible on the backside of terrain features, and behind man-made obstacles, helps conceal communications equipment while still permitting communications.

SIGNAL SECURITY

11-28. EP and signal security are closely related; they are defensive arts based on the same principle. If adversaries do not have access to the essential elements of friendly information (EEFI) of US forces, they are much less effective. The goal of practicing sound EP techniques is to ensure the continued effective use of the electromagnetic spectrum. The goal of signal security is to ensure the enemy cannot exploit the friendly use of the electromagnetic spectrum for communications. Signal security techniques are designed to give commanders confidence in the security of their transmissions. Signal security and EP should be planned based on the enemy's ability to gather intelligence and degrade friendly communications systems.

11-29. Tactical commanders must ensure effective employment of all communications equipment, despite the adversary's concerted efforts to degrade friendly communications to his tactical advantage. Modifying and developing equipment, to make friendly communications less susceptible to adversary exploitation, is an expensive process. Equipment that will solve some EP problems is being developed and fielded. Ultimately, the commander, staff planners, and RTOs are responsible for both security and continued operation of all communications equipment.

EMISSION CONTROL

11-30. The control of friendly electromagnetic emissions is essential to successful defense against the enemy's attempts to destroy or disrupt US communications. Transmitters should be turned on only when needed to accomplish the mission. The enemy intelligence analyst will look for patterns he can turn into usable information. If friendly transmitters are inactive, the enemy has nothing to work with as intelligence. Emission control can be total; for example, the commander may direct radio silence or radio listening silence whenever desired.

11-31. Emission control should be a habitual exercise. Transmissions should be kept to a minimum (20 seconds absolute maximum, 15 seconds maximum preferred) and should contain only mission-critical information. Good emission control makes the use of communications equipment appear random, and is therefore consistent with good EP practices. This technique alone will not eliminate the enemy's ability to find a friendly transmitter; but when combined with other EP techniques, it will make locating a transmitter more difficult.

PREVENTIVE ELECTRONIC PROTECTION TECHNIQUES

11-32. In planning communications, consider the enemy's capabilities to deny the effective use of communications equipment. EP should be planned and applied to force the adversary to commit more jamming, information gathering, and deception resources to a target than it is worth or than he has readily available. EP techniques must also force the enemy to doubt the effectiveness of his jamming and deception efforts.

11-33. RTOs must use preventive EP techniques to safeguard friendly communications from enemy disruption and destruction. Preventive EP techniques include all measures taken to avoid enemy detection, and to deny enemy intelligence analysts useful information. These techniques include EP designed circuits (equipment features) and radio systems installation and operating procedures. Refer to AR 380-5 for the Department of the Army Information Security Program.

11-34. EP designed circuits are in compliance with the MIL-STD for EP. They are built with a focus on technology enhancements, to mitigate the effects of adversary radio electronic combat threats and reduce vulnerabilities to electronic countermeasures.

11-35.   RTOs have little control over the effectiveness of EP designed circuits; therefore, their primary focus is radio systems installation and operating procedures. (Appendix C addresses operations in cold weather, jungle, urban, desert, and nuclear environments.)

11-36.   Incorrect operating procedures can jeopardize the unit's mission, and ultimately increase unit casualties. Communications equipment operators must instinctively use preventive and remedial EP techniques. Maintenance personnel must know that improper modifications to equipment may cause the equipment to develop peculiar characteristics that can be readily identified by the adversary. Commanders and staff must develop plans to ensure the continued use of friendly communications equipment and systems, while also evaluating JSIR reports and after action reports so that appropriate remedial actions can be initiated. FM 7-0 addresses proper training development techniques and is the foundation for developing preventative and EP remedial training.

11-37.   Effective jamming depends on knowing the frequencies and approximate locations of units to be jammed. Using the techniques addressed in the following paragraphs reduces the vulnerability of communications from enemy disruption or destruction; this information must not be disclosed.

MINIMIZING TRANSMISSIONS

11-38.   The most effective preventive EP technique is to minimize both radio transmissions, and transmission times. Although normal day-to-day operations require radio communications, these communications should be kept to the minimum needed to accomplish the mission. Table 11-2 lists the techniques for minimizing transmissions and transmission times.

11-39.   Minimizing transmissions will safeguard radios for critical transmissions. This does not advocate total, continuing radio silence; it advocates minimum transmissions and transmission times.

Table 11-2. Techniques for minimizing transmissions and transmission times

Technique	Description
Ensure all transmissions are necessary.	Analysis of US tactical communications indicates that most communications used in training exercises are explanatory and not directive. Radio communications must never be used as a substitute for complete planning. Tactical radio communications should be used to convey orders and critical information rapidly. Execution of the operation must be inherent in training, planning, ingenuity, teamwork, and established and practiced SOPs. The high volume of radio communications that usually precedes a tactical operation makes the friendly force vulnerable to enemy interception, DF, jamming, and deception.
Note. Even when communications are secure, the volume of radio transmissions can betray an operation, and the enemy can still disrupt or destroy the ability of US forces to communicate.	
Preplan messages before transmitting them.	The RTO should know what he is going to say before beginning a transmission. When the situation and time permit, the message should be written out before beginning the transmission. This minimizes the number of pauses in the transmission and decreases transmission time. It will also help ensure the conciseness of the message. The Joint Interoperability of Tactical Command and Control System (JINTACCS) provides a standard vocabulary that can be used for message planning. JINTACCS voice templates are some of the best tools a RTO can use to minimize transmission time.

Table 11-2. Techniques for minimizing transmissions and transmission times (continued)

Technique	Description
Transmit quickly and precisely.	This is critical when the quality of communications is poor. This minimizes the chances that a radio transmission will have to be repeated. Unnecessary repetition increases transmission time and the enemy's opportunity to intercept US transmissions and thus gain valuable information. When a transmission is necessary, the radio operator should speak in a clear, well-modulated voice, and use proper radiotelephone procedures.
Use equipment capable of data burst transmission.	This is one of the most significant advantages of tactical SATCOM systems. When messages are encoded on a digital entry device for transmission over satellite systems, the transmission time is greatly reduced.
Use an alternate means of communications.	Alternate means of communications, such as cable, wire, or organic Soldiers performing as messengers, can be used to convey necessary directives and information. Other means of communications must be used, when practical.
Use of brevity codes	A brevity code is a code which provides no security but which has as its sole purpose the shortening of messages rather than the concealment of their content. (Refer to FM 1-02.1 for more information.)

Low Power

11-40. Power controls and antennas are closely related. The strength of the signal transmitted by an antenna depends on the strength of the signal delivered to it by the transmitter; the stronger the signal, the farther it travels. A radio communications system must be planned and installed, allowing all stations to communicate with each. In carefully planned and installed communications systems, users can normally operate on low power, thereby decreasing the range, and making it more difficult for the adversary to detect and intercept transmissions. It also reserves high power for penetrating enemy jamming.

RADIO-TELEPHONE OPERATOR PROCEDURES

11-41. The RTO, or Soldier, is essential to the success of preventive EP techniques. The RTO ensures that radio transmissions are minimized and protected; thereby preventing the adversary from intercepting and disrupting or destroying communications based on information detected in the pattern or content of transmissions.

11-42. Many RTOs can be readily identified by certain voice characteristics or overused phrases. The adversary can use these distinguishing characteristics to identify a unit, even though frequencies and net call signs are changed periodically. Strictly adhering to the proper use of procedure words, as outlined in Chapter Twelve or unit SOP, helps keep operator distinguishing characteristics to a minimum. However, this is not enough, as accents and overused phrases must also be kept to a minimum. The adversary must not be able to associate a particular RTO with a particular unit.

11-43. The adversary can gather information based on the pattern, and the content, of radio communications. Therefore, do not develop patterns through hourly radio checks, daily reports at specific times, or any other periodic transmission. Periodic reports should be made by alternate means of communications. Take all reasonable measures to deny information to adversary intelligence analysts.

Authentication

11-44. Authentication must be used in radio systems that do not use secure devices. The adversary has skilled experts, whose sole mission is to enter nets by imitating friendly radio stations. This threat to radio communications can be minimized by the proper use of authentication. Procedures for authentication are found in the supplemental instructions to the SOI. Authentication is required if the user—

- Suspects the adversary is on his net.
- Is challenged by someone to authenticate. (Do not break radio listening silence to do this.)
- Transmits directions or orders that affect the tactical situation, such as change locations, shift fire, or change frequencies.
- Talks about adversary contact, gives an early warning report, or issues a follow-up report. (This rule applies even if he used a brevity list or operations code.)
- Tells a station to go to radio or listening silence, or asks it to break that silence. (Use transmission authentication for this.)
- Transmits to a station that is under radio listening silence. (Use transmission authentication for this.)
- Cancels a message by radio or visual means, and the other station cannot recognize him.
- Resumes transmitting after a long period, or if this is the first transmission.
- Is authorized to transmit a CLASSIFIED message in the clear. (Use transmission authentication for this.)
- Is forced, because of no response by a called station, to send a message in the blind. (Use transmission authentication for this.)

11-45. All instances in which the adversary attempts to deceptively enter nets to insert false information must be reported. The procedures for reporting these incidents are addressed later in this chapter. These procedures are also in the supplemental instructions to the SOI.

Encryption

11-46. Encrypt all EEFI (those items of information the adversary must not be allowed to obtain). A broad, general list of these items of information is contained in the supplemental instructions to the SOI. These items are applicable to most Army units engaged in training exercises or tactical operations. The list supports the Army self-monitoring program, and is not totally encompassing. Individual units should develop a more specific EEFI list to be included in unit OPORDS, OPLANS, and field SOPs. These items of information must be encrypted manually or electronically before transmission. Manual encryption is accomplished by using approved operations codes. Electronic encryption is accomplished using COMSEC devices such as the KG-84, KG-95, KY-57/58, KY-90, KY-99, and KY-100 or ANCD/SKL. Manual and electronic encryption does not need to be used together, as either method will protect EEFI from enemy exploitation.

Key Distribution

11-47. Key distribution is critical in achieving secure transmissions. Commanders must ensure these procedures are established in the units SOP. Only the requesting unit's COMSEC custodian, with valid COMSEC account (and requirement), is authorized to order these keys.

11-48. TB 11-5820-890-12 and TM 11-5820-1130-12&P expound upon the receipt of OTAR material and TB 380-41 provides more information on the procedures for safeguarding, accounting, and the supply control of COMSEC material such as COMSEC material distribution.

EQUIPMENT AND COMMUNICATIONS ENHANCEMENTS

11-49. Equipment enhancements can be used to reduce the vulnerability of friendly communications to hostile exploitations. FH is particularly useful in lessening the effects of adversary communications jamming, and in denying friendly position location data to the enemy.

11-50. Adaptive antenna techniques are designed to achieve more survivable communications systems. These techniques are typically coupled with spread spectrum waveforms, combining FH with pseudo-noise coding.

11-51. Spread spectrum techniques suppress interference by other users (hostile or friendly), to provide multiple access (user sharing), and to eliminate multi-path interference (self-jamming caused by a delayed signal). The transmitted intelligence is deliberately spread across a very wide frequency band in the operating spectrum, so it becomes hard to detect from normal noise levels. EPLRS and JTIDS use spread spectrum techniques.

11-52. Adjustable power automatically limits the radiated power to a level sufficient for effective communications, thereby reducing the electronic signature of the subscriber.

11-53. The FHMUX and high power broadband vehicular whip antennas are available for use to enhance communications. The FHMUX is an antenna multiplexer used with SINCGARS in both stationary and mobile operations. This multiplexer will allow up to four SINCGARS to transmit and receive through one VHF-FM broadband antenna (OE-254 or high-power broadband vehicular whip antenna) while operating in the FH mode, non-hopping mode, or a combination of both. Using one antenna (instead of up to four) will reduce visual and electronic profiles of CPs. Also, emplacement and displacement times will be greatly reduced.

REMEDIAL EP TECHNIQUES

11-54. Remedial EP techniques that help reduce the effectiveness of enemy efforts to jam US radio nets are—

- Identify jamming signals.
- Determine if the interference is obvious or subtle jamming.
- Recognize jamming and interference by:
    - Determining whether the interference is internal or external to the radio.
    - Determining whether the interference is jamming or unintentional.
    - Reporting jamming and interference incidents.
- Overcome jamming and interference by adhering to the following techniques:
    - Continue to operate.
    - Improve the signal-to-jamming ratio.
    - Adjust the receiver.
    - Increase the transmitter power output.
    - Adjust or change the antenna.
    - Establish a wireless network extension station.
    - Relocate the antenna.
    - Use an alternate route for communications.
    - Change the frequencies.
    - Acquire another satellite.

JAMMING SIGNALS

11-55. Jamming is an effective way for the enemy to disrupt friendly communications. An adversary only needs a transmitter tuned to a US frequency, with enough power to override friendly signals, to jam US systems. Jammers operate against receivers, not transmitters. The two modes of jamming are spot and barrage jamming. Spot jamming is concentrated power directed toward one channel or frequency. Barrage jamming is power spread over several frequencies or channels at the same time. It is important to recognize jamming, but it can be difficult to detect.

Obvious Jamming

11-56. Obvious jamming is normally simple to detect. When experiencing jamming, it is more important to recognize and overcome the incident than to identify it formally. Table 11-3 lists some common jamming signals.

Table 11-3. Common jamming signals

Signal	Description
Random Noise	Synthetic radio noise. It is indiscriminate in amplitude and frequency. It is similar to normal background noise, and can be used to degrade all types of signals. Operators often mistake it for receiver or atmospheric noise, and fail to take appropriate EP actions.
Stepped Tones	Tones transmitted in increasing and decreasing pitch. They resemble the sound of bagpipes. Stepped tones are normally used against SC AM or FM voice circuits.
Spark	Easily produced and is one of the most effective jamming signals. Bursts are of short duration and high intensity; they are repeated at a rapid rate. This signal is effective in disrupting all types of radio communications.
Gulls	Generated by a quick rise and slow fall of a variable RF, and are similar to the cry of a sea gull. It produces a nuisance effect and is very effective against voice radio communications.
Random Pulse	Pulses of varying amplitude, duration, and rate are generated and transmitted. They are used to disrupt teletypewriter, radar, and all types of data transmission systems.
Wobbler	A single frequency, modulated by a low and slowly varying tone. The result is a howling sound that causes a nuisance effect on voice radio communications.
Recorded Sounds	Any audible sound, especially of a variable nature, can be used to distract radio operators and disrupt communications. Music, screams, applause, whistles, machinery noise, and laughter are examples.
Preamble Jamming	A tone resembling the synchronization preamble of the speech security equipment is broadcast over the operating frequency of secure radio sets. Results in all radios being locked in the receive mode. It is especially effective when employed against radio nets using speech security devices.

Subtle Jamming

11-57. Subtle jamming is not obvious, as no sound is heard from the receivers. Although everything appears normal to the RTO, the receiver cannot receive an incoming friendly signal. Often, users assume their radios are malfunctioning, instead of recognizing subtle jamming for what it is.

RECOGNIZING JAMMING

11-58. RTOs must be able to recognize jamming. This is not always an easy task, as interference can be internal and external. If the interference or suspected jamming remains, after grounding or disconnecting the antenna, the disturbance is most likely internal and caused by a malfunction of the radio. Maintenance personnel should be contacted to repair it. If the interference or suspected jamming can be eliminated or substantially reduced by grounding the radio equipment or disconnecting the receiver antenna, the source of the disturbance is most likely external to the radio. External interference must be checked further for enemy jamming or unintentional interference.

11-59. Interference may be caused by sources having nothing to do with enemy jamming. Unintentional interference may be caused by—
- Other radios (friendly and enemy).
- Other electronic or electric/electromechanical equipment.
- Atmospheric conditions.
- Malfunction of the radio.
- A combination of any of the above.

11-60.   Unintentional interference normally travels only a short distance; a search of the immediate area may reveal its source. Moving the receiving antenna short distances may cause noticeable variations in the strength of the interfering signal. Conversely, little or no variation normally indicates enemy jamming. Regardless of the source, actions must be taken to reduce the effect of interference on friendly communications.

11-61.   The enemy can use powerful unmodulated or noise modulated jamming signals. Unmodulated jamming signals are characterized by a lack of noise. Noise modulated jamming signals are characterized by obvious interference noise.

11-62.   In all cases, suspected enemy jamming and any unidentified or unintentional interference that disrupts the ability of US forces to communicate must be reported. This applies even if the radio operator is able to overcome the effects of the jamming or interference. The JSIR report is the format used when reporting this information. Instructions for submitting a JSIR report are addressed later in this chapter. As it applies to remedial EP techniques, the information in the JSIR report provided to higher headquarters can be used to destroy the enemy jamming efforts or take other action to the benefit of US forces.

OVERCOMING JAMMING

11-63.   The enemy constantly strives to perfect and use new and more confusing forms of jamming. RTOs must be increasingly alert to the possibility of jamming. Training and experience are the most important tools operators have to determine when a particular signal is a jamming signal. Exposure to the effects of jamming in training, or actual situations, is invaluable. The ability to recognize jamming is important, as jamming is a problem that requires action. The following paragraphs address the actions to take if adversary jamming is detected. If any of the actions taken alleviate the jamming problem, simply continue normal operations and submit a JSIR report to higher headquarters.

Continue to Operate

11-64.   Adversary jamming usually involves a period of jamming followed by a brief listening period. Operator activity during this short period of time will tell the adversary how effective his jamming has been. If the operation is continuing in a normal manner, as it was before the jamming began, the adversary will assume that his jamming has not been particularly effective. On the other hand, if he hears a discussion of the problem on the air or if the operation has been shut down entirely, the enemy may assume that his jamming has been effective. Because the adversary jammer is monitoring operation this way, unless otherwise ordered, never terminate operations or in any way disclose to the enemy that the radio is being adversely affected. This means normal operations should continue even when degraded by jamming.

Improve the Signal-to-Jamming Ratio

11-65.   The signal-to-jamming ratio is the relative strength of the desired signal to the jamming signal at the receiver. Signal refers to the signal being received. Jamming refers to the hostile or unidentified interference being received. It is always best to have a signal-to-jamming ratio in which the desired signal is stronger than the jamming signal. In this situation, the desired signal cannot be significantly degraded by the jamming signal. To improve the signal-to-jamming ratio operators and signal leaders can consider the following—

- **Increase the transmitter power output**. To increase the power output at the time of jamming, the transmitter must be set on something less than full power when jamming begins. Using low power as a preventive EP technique depends on the enemy not being able to detect radio transmissions. Once the enemy begins jamming the radios, the threat of being detected becomes obvious. Use the reserve power on the terrestrial LOS radios to override the enemy's jamming signal.
- **Adjust or change the antenna**. When jamming is experienced, the radio operator should ensure the antenna is optimally adjusted to receive the desired incoming signal. Specific methods that

apply to a particular radio set are in the appropriate operator's manual. Depending on the antenna, some methods include—

- Reorienting the antenna.
- Changing the antenna polarization. (Must be done by all stations.)
- Installing an antenna with a longer range.

- **Establish a wireless network extension station**. This can increase the range and power of a signal between two or more radio stations. Depending on the situation and available resources, this may be a viable method to improve the signal-to-jamming ratio.

- **Relocate the antenna**. Frequently, the signal-to-jamming ratio may be improved by relocating the antenna and associated radio set affected by the jamming or unidentified interference. This may mean moving it a few meters or several hundred meters. It is best to relocate the antenna and associated radio set, so there is a terrain feature between them and any suspected enemy jamming location.

- **Use an alternate route for communications**. In some instances, enemy jamming will prevent friendly forces from communicating with another radio station. If radio communications have been degraded between two radio stations that must communicate, another radio station or route of communications may be used as a relay between the two radio stations.

- **Change frequencies**. If a communications net cannot overcome enemy jamming using the above measures, the commander (or designated representative) may direct the net to be switched to an alternate or spare frequency. If practical, dummy stations can continue to operate on the frequency being jammed, to mask the change to an alternate frequency; this action must be preplanned and well coordinated. During enemy jamming, it may be difficult to coordinate a change of frequency. All RTOs must know when, and under what circumstances, they are to switch to an alternate or spare frequency. If this is not done smoothly, the enemy may discover what is happening, and try to degrade communications on the new frequency.

ELECTRONIC WARFARE FOR SINGLE-CHANNEL TACTICAL SATELLITE

11-66. SC TACSAT communications is an important element of the C2 system. Parts of the enemy's resources are directed against the satellite system through EW. How vulnerable we are to enemy EW and the success of our actions to deny the enemy success in EW efforts depends on our equipment and our signal personnel.

11-67. SC TACSAT communications will be high on the enemy's target list. Shortly after tactical communications is placed in operation, the enemy will compile data on the satellite. This data will most likely include—

- Data indicating the satellite's orbit and location.
- Information on frequency, bandwidth, and modulation used in the satellite.
- The amount, type, and frequency of traffic relayed by the satellite.

11-68. With the satellite relay located, the primary enemy threat then is directed toward locating ground stations through radio DF. Due to the highly directional antennas used with super high frequency/EHF SC TACSAT communications radios, there is a low probability of intercept and DF. However, a satellite based intercept station orbiting near our satellites can be successful. In this case, the analysis effort can be done by the enemy on his home ground, far from the AO.

11-69. Because of the enemy's massive computer support SC TACSAT communications stations will hide very little from the enemy. Even without ground station locations, jamming can be directed towards the satellites. When this is occurs, SC TACSAT communications nets working through the satellite are operating in a "stressed" mode. Jamming signals directed toward the satellite can originate far from the battlefield. Due to the directional antennas and frequencies used, jamming directed toward ground stations must come from nearby. Besides jamming, the enemy may attempt deception from either the ground or his own satellites. The enemy may attempt to insert false or misleading information and may also establish dummy nets operating through our satellites to cause confusion. In stability operations however, there is a reduced electronic threat.

DEFENSIVE ELECTRONIC WARFARE

11-70. TACSAT communications must operate within the environment just described. To do this, it is necessary to use available anti-jamming equipment and sound countermeasures. Communications discipline, security, and training underlie ECCM. COMSEC techniques give the commander confidence in the security of his communications. ECCM equipment and techniques provide confidence in the continued operation of TACSAT communications in a hostile EW or stressed environment. Particularly in SC TACSAT communications, the two are closely related techniques serving an ECCM role.

11-71. COMSEC techniques protect the transmitted information. Physical security safeguards COMSEC materiel and information from access or observation by unauthorized personnel using physical means. TRANSEC protects transmissions from hostile interception and exploitation. COMSEC and TRANSEC equipment protects most circuits. However, some SC TACSAT orderwires may not be secure. Technical discussions between operators can contain information important to the enemy. The nature of any mission gives the enemy access to critical information about commanders, organizations, and locations of headquarters. Although revealed casually on the job, this information is sensitive and must be protected.

11-72. ECCM techniques protect against enemy attempts to detect, deceive, or destroy friendly communications. Changing frequency can defeat jamming. This requires the jammer to determine the new frequency and move to it. Meanwhile, the frequency can again be changed. This is the principle behind FH.

11-73. Since it takes about 0.25 seconds for the earth station satellite-earth station trip, FH four times per second denies the jammer access to the satellite to earth link. FH at this rate must rely on automated equipment. FH at rates between 4 per second and 75 per second effectively avoids intercept and jamming when the enemy can receive only the downlink. With these low rates, bandwidth is still minimal while providing secure communications. FH forces the jammer to spread his energy (broadband jamming). This reduces the jammer noise density on any one channel.

11-74. Wideband spread spectrum modulation is another effective anti-jamming technique. With this technique, the information transmitted is added to a pseudorandom noise code and is used to modulate the SC TACSAT terminal transmitter. At the receiving end, an identical noise generator synchronized to the transmitter is used. It generates the same noise code as the one at the transmitter to cancel the noise signal from the incoming signal. Thus, only the transmitted information remains.

11-75. The spread spectrum signal can occupy the entire bandwidth of the satellite at the same time with several other spread spectrum signals. Each signal must have a different pseudorandom noise code. The noise code looks the same to the jammer whether or not it is carrying intelligence. This forces the jammer to spread his energy throughout the entire bandwidth of the random noise. This results in a reduced jamming noise density. The jammer has no knowledge of whether the jamming is effective.

ELECTROMAGNETIC COMPATIBILITY

11-76. An electromagnetic compatibility occurs when all equipment (radios, radars, generators) and vehicles (ignition systems) operate without interference from each other. With SC TACSAT communications terminals, a source of interference is solar weather (to include solar flares, solar winds, geomagnetic storms, and solar radiation storms). However, factors such as location and antenna orientation can be controlled to eliminate this source of noise. For each piece of equipment, use proper grounding techniques and follow safety considerations. When SC TACSAT communications terminals and other sets must be collocated, use a plan that prevents antennas from shooting directly into one another. Maintaining an adequate distance between antennas reduces mutual interference.

11-77. Desensitization is the most common interference problem. This reduces receiver sensitivity caused by signals from nearby transmitters. The Electromagnetic compatibility must be included in the plans for siting a SC TACSAT communications station. An electromagnetic pulse (EMP) is a threat to all sophisticated electronic systems.

COUNTER REMOTE CONTROL IMPROVISED EXPLOSIVE DEVICE WARFARE

11-78.   Counter Remote Control Improvised Explosive Device Warfare (CREW) provides the operational capability to prevent and/or defeat improvised explosive device (IED) detonation ambushes that are pervasive threats throughout an AOR. CREW employs a spiral development approach to allow for rapid fielding of incremental CREW capabilities. CREW acts a radio-frequency jammer to preempt the detonation of remote control IEDs by disrupting the radio signal.

11-79.   CREW-1 produced and fielded the Warlock Family of Systems. Crew-1 systems are in various configurations and varied levels of performance. CREW-1 systems target specific RF dependent technologies. CREW-1 Warlock Systems are—

- Warlock-Red (Figure 11-2).
- Warlock-Green.
- Warlock-IED Counter-Measure Equipment.
- Warlock-Self Screening Vehicle Jammer.
- Warlock-Blue Wearable & Vehicle Mounted.

Figure 11-2. Warlock-red

JOINT SPECTRUM INTERFERENCE RESOLUTION REPORTING

11-80.   JSIR addresses EMI incidents and EA affecting the DOD. The JSIR objective is to report and assist in resolving EA and persistent, recurring interference. Resolution is at the lowest possible level, using organic assets. Incidents that cannot be resolved locally are referred up the chain of command with resolution attempted at each level.

11-81.   Chairman Joint Chiefs of Staff Instruction (CJCSI) 3320.02A directs DOD components to resolve RF interference at the lowest possible level within the chain of command. To accomplish this, the Army established the Army interference resolution program (AIRP).

ARMY INTERFERENCE RESOLUTION PROGRAM

11-82.   The AIRP revolves around four functions: DF, signal monitoring, signal analysis, and transportability/mobility. These functions are described in Table 11-4. Refer to AR 5-12 for additional information on the AIRP.

Table 11-4. Army interference resolution program functions

Function	Description
DF	Is often the key to locating the source of interference, and is an integral part of resolving and analyzing incidents and problems. The degree of accuracy depends upon the environment and frequency band.
Signal Monitoring	Or spectrum surveillance incorporates a frequency spectrum analyzer or surveillance receiver, covering all spectrum bands of use. These systems perform real-time evaluation of spectrum usage and interference in a specific area.
Signal Analysis	Analysis of DF and monitoring data is required to determine the source of interference and misuse of the spectrum.
Transportabi lity/Mobility	Degree, circumstances, and geographic location of the types of interference incidents and problems will determine transportability and mobility requirements. Mobile/Transportable DF and monitoring equipment is a requirement for tactical units and for incidents not necessarily confined to a specific geographical area. Man portable equipment should be considered for certain instances and conditions, as defined in unit SOPs. Fixed equipment would be required for those areas that require real-time solutions in a defined geographical area.

INTERFERENCE RESOLUTION

11-83. Corps and division frequency managers are the coordinating authorities for regional and local interference resolution. The impact of each interference incident is unique, and no standard procedure can be established that will guarantee resolution in every case. However, a logical step-by-step approach will reduce time and cost in resolving interference situations. Figure 11-3 is a logical flow diagram for instances when an Army unit is the victim of interference in a tactical operation. Figure 11-4 shows a flow diagram for interference, when the Army unit is the source of the interference.

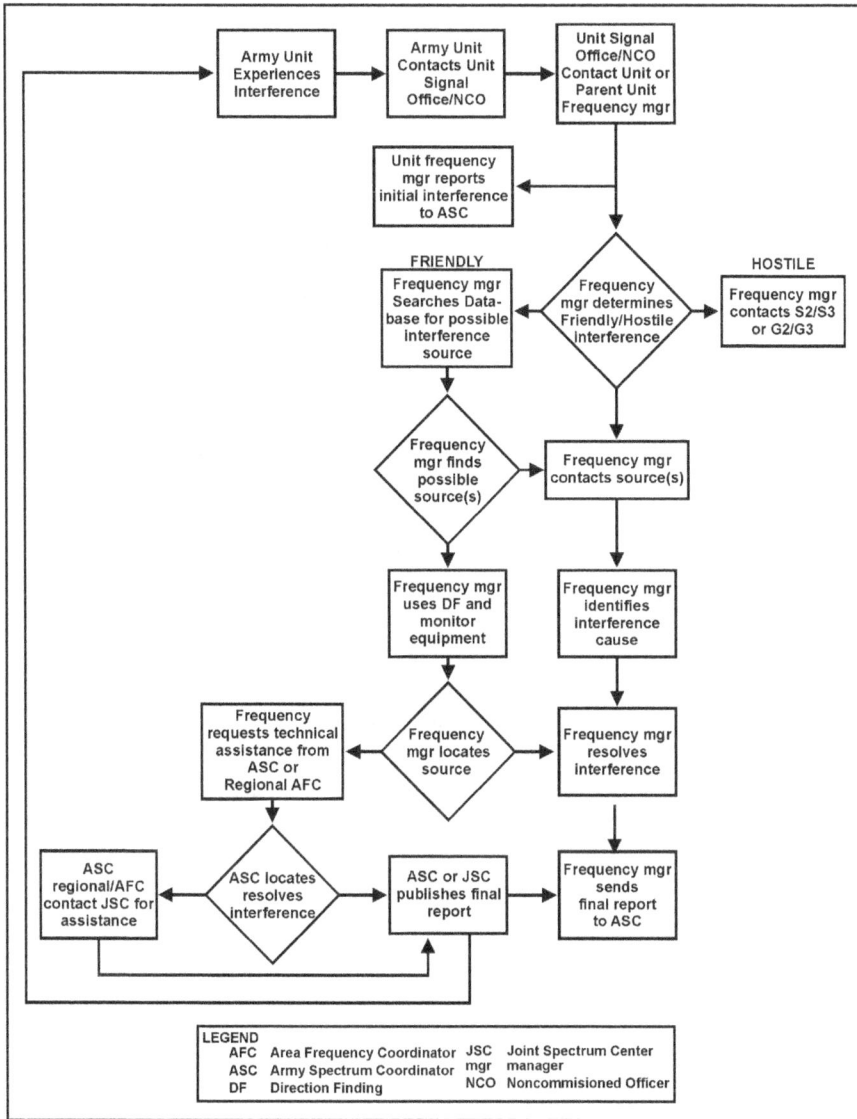

Figure 11-3. Interference resolution (Army victim)

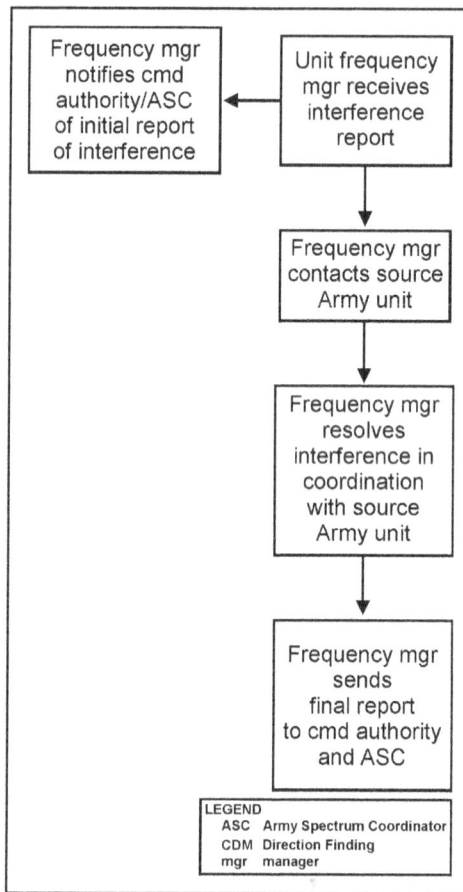

Figure 11-4. Interference resolution (Army source)

Reporting Procedure

11-84. All EMI incidents must be reported through the proper channels. All reports of suspected hostile interference are submitted via secure means. The report should not be held up due to information not being readily available; use follow-up reports to provide additional information, as it becomes available.

11-85. The equipment operator experiencing the interference incident forwards the initial JSIR report through the chain of command to the unit operations center. An attempt to resolve the EMI problem at the lowest possible level will be conducted before submitting JSIR reports to higher headquarters.

11-86. The Joint Spectrum Management System/Spectrum XXI programs should be used to submit the report electronically. The sender will classify the report by evaluating the security sensitivity of the interference on the affected system, and by considering the classification of the text comments. Table 11-5 is a guide for JSIR security classification.

11-87. The JSIR report will be assigned precedence consistent with the urgency of the reported situation. Use ROUTINE or PRIORITY precedence, unless the organization originating the report believes the incident is hazardous to military operations. For this incident, use IMMEDIATE precedence.

11-88. Each Army unit must submit reports through its chain of command, up to the major, or combatant command, or GCC level, and to the US Army Communications-Electronic Services Office. Information copies of all incident reports should be sent to Joint Spectrum Center for inclusion in the JSIR database.

Table 11-5. JSIR security classification guide

Information Revealing	Security Classification
The specific identification of an unfriendly platform or location, by country or coordinates, as the source of interference or EA.	SECRET (S)
Specific susceptibility or vulnerability of US electronic equipment/systems.	SECRET (S)
Parametric data of classified US electronic equipment.	In accordance with the classification guide of the affected equipment.
Suspected interference from unidentified sources while operating in or near hostile countries.	SECRET (S)
Interference to US electromagnetic equipment/systems caused by EA exercises in foreign nations.	CONFIDENTIAL
Suspected interference from friendly sources.	UNCLASSIFIED (U) or SECRET (S), if specific equipment vulnerability is revealed.
Information referring to JSIR; stating that JSIR analyses are a function of the Joint Spectrum Center.	UNCLASSIFIED (U)

Joint Spectrum Interference Resolution Report Content

11-89. Table 11-6 shows the minimum information requirements for the JSIR. The message subject line should indicate whether the report is initial, follow-up, or final.

Table 11-6. JSIR information requirements

Item Number	Data Input
1	Frequencies affected by the interference.
2	Locations of systems experiencing the interference.
3	The affected system name, nomenclature, manufacturer (with model number), or other system description. If available, include the equipment characteristics of the victim receiver, such as bandwidth, antenna type, and antenna size.
4	The operating mode of the affected system. If applicable, include the following: frequency agile, pulse Doppler, search, and upper and lower sidebands.
5	The characteristics of the interference (noise, pulsed, continuous, intermittent, frequency, or bandwidth).
6	The description of the interference effects on victim performance (reduced range, false targets, reduced intelligibility, or data errors).
7	Enter the dates and times the interference occurred. Indicate whether the duration of the interference is continuous or intermittent, the approximate repetition rate of the interference, and whether the amplitude of the interference is varying or constant. Indicate if the interference is occurring at a regular or irregular time of day, and if the occurrence of the interference coincides with any ongoing local activity.
8	The location of possible interference sources (coordinates or line of bearing, if known; otherwise, state as unknown).

Table 11-6. JSIR information requirements (continued)

Item Number	Data Input
9	A listing of other units affected by the interference (if known) and their location or distance, and bearing from the reporting site.
10	A clear and concise narrative summary of what is known about the interference, and any local actions that have been taken to resolve the problem. The operator is encouraged to provide any other information, based on observation or estimation that is pertinent in the technical or operational analysis of the incident. Identify whether the information being furnished is based on actual observation/measurement or is being estimated. Avoid the use of Army or program jargon and acronyms.
11	Reference message traffic that is related to the interference problem being reported. Include the message date-time group, originator, action addressees, and subject line.
12	Indicate whether the problem has been identified or resolved.
13	Indicate if JSIR technical assistance is desired or anticipated.
14	Point of contact information, including name, unit, and contact phone numbers.

Chapter 12

Radio Operating Procedures

Using proper radio procedures can make the difference in time and security when operating on C2 nets. This chapter addresses the proper way to pronounce letters and numbers when sending messages over a radio as well as the proper procedures for opening and closing a radio net.

PHONETIC ALPHABET

12-1. When radio operators are communicating over the radio they will use the phonetic alphabet outlined in Table 12-1 to pronounce individual letters of the alphabet.

Table 12-1. Phonetic alphabet

LETTER	WORD	PRONUNCIATION
A	ALPHA	AL FAH
B	BRAVO	BRAH VOH
C	CHARLIE	CHAR LEE OR SHAR LEE
D	DELTA	DELL TAH
E	ECHO	ECH OH
F	FOXTROT	FOKS TROT
G	GOLF	GOLF
H	HOTEL	HOH TELL
I	INDIA	IN DEE AH
J	JULIETT	JEW LEE ETT
K	KILO	KEY LOH
L	LIMA	LEE MAH
M	MIKE	MIKE
N	NOVEMBER	NO VEM BER
O	OSCAR	OSS CAH
P	PAPA	PAH PAH
Q	QUEBEC	KEH BECK
R	ROMEO	ROW ME OH
S	SIERRA	SEE AIR RAH
T	TANGO	TANG GO
U	UNIFORM	YOU NEE FORM OR OO NEE FORM
V	VICTOR	VIC TAH
W	WISKEY	WISS KEY
X	XRAY	ECKS RAY
Y	YANKEE	YANG KEY
Z	ZULU	ZOO LOO

NUMERICAL PRONUNCIATION

12-2. To distinguish numerals from words similarly pronounced, the proword "FIGURES" may be used preceding such numbers.

12-3. Table 12-2 outlines the pronunciation of how numerals will be transmitted by radio.

Table 12-2. Numerical pronunciation

NUMERAL	SPOKEN AS
0	ZE-RO
1	WUN
2	TOO
3	TREE
4	FOW-ER
5	FIFE
6	SIX
7	SEV-EN
8	AIT
9	NIN-ER

12-4. Numbers will be transmitted digit by digit except that exact multiples of thousands may be spoken as such (refer to Table 12-3). However, there are special cases, such as anti-air warfare reporting procedures, when the normal pronunciation of numerals is prescribed for example, 17 would then be "seventeen."

Table 12-3. Numerals in combinations

NUMBERAL	SPOKEN AS
44	FOW-ER, FOW-ER
90	NIN-ER, ZE-RO
136	WUN, TREE, SIX
TIME 1200	WUN, TOO, ZE-RO, ZE-RO
1748	WUN, FOW-ER, SEV-EN, AIT
7000	SEV-EN, TOU-SAND
16000	WUN, SIX, TOU-SAND
812681	AIT, WUN, TOO, SIX, AIT, WUN

12-5. The figure "ZERO" will be written as "0," the figure "ONE" will be written as "1" and the letter "ZULU" will be written as "Z". Difficult words may be spelled out phonetically but abbreviations and isolated letters should be phoneticized without the proword "I SPELL".

> *Note.* Any abbreviated words used in the message must be transmitted phonetically, for example, 1ˢᵗ is sent as ONE SIERRA TANGO, or headquarters (HQ) as HOTEL QUEBEC.

PROCEDURE WORDS

12-6. Table 12-4 outlines proper procedure words (often called prowords) that should be used during radio transmissions and their meanings. Prowords are words or phrases limited to radio telephone procedures used to facilitate communication by conveying information in a condensed form.

Table 12-4. Prowords listed alphabetically

PROWORD	MEANING
ACKNOWLEDGE	A directive from the originator requiring the addressee (s) to advise the originator that his communication has been received and understood. This term is normally included in the electronic transmission of orders to ensure the receiving station or person confirms the receipt of the orders.
ALL AFTER	The portion of the message to which I have referenced is all that which follows.
ALL BEFORE	The portion of the message to which I have reference is all that proceeds.
AUTHENTICATE	The station called is to reply to the challenge which follows.
AUTHENTICATION IS	The transmission authentication of this message is.
BREAK	I hereby indicated the separation of the text from other portions of the message.
CLEAR	To eliminate transmission on a net in order to allow a higher-precedence transmission to occur.
CORRECT	You are correct, or what you have transmitted is correct.
CORRECTION	An error has been made in this transmission. Transmission will continue with the last word correctly transmitted.
DISREGARD THIS TRANSMISSION-OUT	This transmission is in error. Disregard it. (The proword shall not be used to cancel any message that has been completely transmitted and for which receipt or acknowledgement has been received.)
DO NOT ANSWER	Stations called are not to answer this call, receipt for this message, or otherwise to transmit in connection with this transmission. When this proword is employed, the transmission shall be ended with the proword "OUT".
EXEMPT	The addressees immediately following are exempted from the collective call.
FIGURES	Numerals or numbers follow. (Optional)
FLASH	Precedence FLASH. Reserved for initial enemy contact reports on special operational combat traffic originated by specifically designated high commanders of units directly affected. This traffic is SHORT reports of emergency situations of vital proportion. Handling is as fast as possible with an objective time of 10 minutes or less.
FROM	The originator of this message is indicated by the address designator immediately following.
GROUPS	This message contains numbers of groups indicated.
I AUTHENTICATE	The group that follows it is the reply to your challenge to authenticate.
IMMEDIATE	Precedence IMMEDIATE. Reserved for messages relating to situations which gravely affect the security of national/multinational forces of populace, and which require immediate delivery.
INFO	The addressees immediately following are addressed for information.
I READ BACK	The following is my response to your instructions to read back.
I SAY AGAIN	I am repeating transmission or portion indicated.
I SPELL	I shall spell the next word phonetically.

Table 12-4. Prowords listed alphabetically (continued)

PROWORD	MEANING
I VERIFY	That which follows has been verified at your request and is repeated. (To be used as a reply to verify information.)
MESSAGE	A message which requires recording is about to follow. (Transmitted immediately after the call.)
MORE TO FOLLOW	Transmitting station has additional traffic for the receiving station.
OUT	This is the end of my transmission to you and no answer is required or expected. (Since OVER and OUT have opposite meanings, they are never used together.)
OVER	This is the end of my transmission to you and a response is necessary. Go ahead; transmit.
PRIORITY	Precedence PRIORITY. Reserved for important messages which must have precedence over routine traffic. This is the highest precedence which normally may be assigned to a message of administrative nature.
READ BACK	Repeat this entire transmission back to me exactly as received.
RELAY (TO)	Transmit this message to all addressee (or addresses immediately following this proword). The address component is mandatory when this proword is used.
ROGER	I have received your last transmission satisfactorily.
ROUTINE	Precedence ROUTINE. Reserved for all types of messages which are not of sufficient urgency to justify a higher precedence, but must be delivered to the addressee without delay.
SAY AGAIN	Repeat all of your last transmission. (Followed by identification data means to repeat after the portion indicated.
SILENCE	"Cease Transmission Immediately." Silence will be maintained until lifted. (Transmission imposing silence must be authenticated.)
SILENCE LIFTED	Silence is lifted. (When authentication system is in force the transmission silence is to be authenticated.)
SPEAK SLOWER	Your transmission is at too fast of a speed. Reduce speed of transmission.
THIS IS	This transmission is from the station whose designator immediately follows.
TIME	That which immediately follows is the time or date/time group of the message.
SILENCE	"Cease Transmission Immediately." Silence will be maintained until lifted. (Transmission imposing silence must be authenticated.)
TO	The addressee(s) immediately following is (are) addressed for action.
UNKNOWN STATION	The identity of the station with whom I am attempting to establish communications is unknown.
VERIFY	Verify the entire message (or portion indicated) with the originator and send correct version. (To be issued only at the discretion of the addressee to which the questioned message was directed.)
WAIT	I must pause for a few seconds.

Table 12-4. Prowords listed alphabetically (continued)

PROWORD	MEANING
WAIT OUT	I must pause for longer than a few seconds.
WILCO	I have received your signal, understand it and will comply. (To be used only by the addressee. Since the meaning of ROGER is included in that of WILCO, the two prowords are never used together)
WORD AFTER	The word of the message to which I have reference is that which follows...
WORD BEFORE	The word of the message to which I have reference is that which proceeds...
WORD TWICE	Communication is difficult. Transmit (ring) each phrase (or each code group) twice. This procedure word may be used as an order, request, or as information.
WRONG	Your last transmission was incorrect. The correct version is...

RADIO CALL PROCEDURES

12-7. A preliminary call will be transmitted when the sending station wishes to know if the receiving station is ready to receive a message. When communications reception is good and contact has been continuous, a preliminary call is optional. The following is an example of a preliminary call—

- A1D THIS IS B6T, OVER.
- B6T THIS IS A1D, OVER.
- A1D THIS IS B6T (sends message), OVER.
- B6T THIS IS A1D, ROGER OUT.

Note. For more information on radio call signs and procedures refer to Allied Communications Publication 121 and 125.

OPENING A RADIO NET

12-8. During radio net calls, the last letter of the call sign determines the answering order. The stations in a net respond alphabetically, for example, A3D will answer before A2W and A2E will answer before BIF. If two stations in a net have the same last letter, for instance, A1D and A2D, then the answering order will be determined by numerical sequence, with the lower number A1D answering first.

12-9. The following is an example of a secure voice net opening by the NCS and several distant stations—

- NET THIS NCS, OVER.
- NCS THIS IS A1D, OVER.
- NCS THIS IS A2D, OVER.
- NCS THIS A2E, OVER.
- NET THIS IS NCS, OUT (IF THE NCS HAS NO TRAFFIC).

RADIO CHECKS

12-10. To minimize transmission time, use radio checks sparingly or by unit SOP. The following is an example of a radio check with the NCS—

- NET THIS IS NCS, RADIO CHECK OVER.
- NCS THIS IS A1D, ROGER OUT.
- NCS THIS IS A2D, WEAK READABLE OVER (A2D is receiving the NCS's signal weak).
- NCS THIS IS A2E, ROGER OUT.
- NET THIS IS NCS, ROGER OUT.

STATION ENTERING A NET ALREADY ESTABLISHED

12-11.   The following is an example of how a radio station would enter a net after the net was opened and the station was unable to answer and now wants to report into the net (NCS)—

- NCS THIS B4G, REPORTING INTO THE NET OVER.
- B4G THIS NCS, AUTHENTICATE OVER.
- NCS THIS B4G, I AUTHENTICATE (B4G authenticates) OVER.
- B4G THIS IS NCS, I AUTHENTICATE (NCS authenticates) OVER.
- NCS THIS IS B4G, ROGER OUT.

Note. Authentication is a security measure designed to protect a communications system against acceptance of a fraudulent transmission or simulation by establishing the validity of a transmission, message, or originator.

STATION REQUESTING TO LEAVE A NET

12-12.   The following is an example of a radio station requesting permission to leave a net from the NCS of the net—

- NCS THIS A24, REQUEST PERMISSION TO CLOSE DOWN (OR LEAVE NET), OVER.
- A24 THIS IS NCS, ROGER OUT.

CLOSING A SECURE VOICE NET

12-13.   The following is an example of a NCS closing a secure voice radio net. Authentication can be used for a non secure net.

- NET THIS IS NCS, CLOSE DOWN, OVER.
- NCS THIS A1D, ROGER OUT.
- NCS THIS A2D, ROGER OUT.
- NCS THIS B2D, ROGER OUT.

Note. For more information on NCS radio procedures refer to TM 11-5820-890-10-5 and TM 11-5820-890-10-8.

Appendix A

FM Radio Networks

Units from battalion to theater establish FM radio nets, for example C2, fires net, A&L, and O&I nets to execute on the move combat operations. Commanders may establish other networks in addition to these to enhance mission accomplishment. The lack of sufficient SC TACSAT frequency resources, SC radio systems density and the need for radio wireless network extension capability all validate the need for FM networks. This appendix addresses FM networks.

COMMAND AND CONTROL NETWORKS

A-1. C2 networks are found in all Army units. The units establish internal C2 networks, and are subscribers in at least one other network. SINCGARS is the primary means of short range communications in secure C2 voice networks. The C2 net is given the highest installation priority.

A-2. Table A-1 is an example of division networks. The C2 networks shown merely serve as a guide for establishing radio networks. The actual networks established depend on the existing situation, command guidance, and equipment available. Figure A-1 is an example of typical subscribers for a division C2 FM network.

Note. Subscribers in a C2 network are members of that echelon and the next senior echelon C2 network. When necessary, wireless network extension teams are used to overcome communications obstacles between higher and lower units.

Table A-1. Example of division C2 FM networks

Net Stations	Command (CMD) Operations Net	O&I Net	Sustainment Operations Net	A&L Net
Commander (CDR)	X	X	X	X
Assistant CDR	X			
OP G-3	X	X	X	X
G-2		X		X
TAC CP G-3	X		X	X
TAC CP G-2		X		
TAC CP G-6	X	X	X	X
Subordinate brigade CP	X	X	X	X
Brigade support battalion	X		X	
Reconnaissance battalion	X	X	X	X
Aviation units	X	X	X	X
Engineer unit	X	X	X	X

Table A-1. Example of division C2 FM networks (continued)

Net Stations	Command Operations Net	O&I Net	Sustainment Operations Net	A&L Net
Military intelligence unit	X	X		
ADA unit	X	X		X
Artillery units	X	X	X	X
Military police	X		X	
Sustainment operations center	X		X	X
Division Signal company	X	X	X	
Liaison officer	X			
Long range reconnaissance detachment		X		

TACTICAL THEATER SIGNAL BRIGADES

A-3. A tactical theater signal brigade (TTSB) provides the Army and joint forces with an agile, expeditionary-capable signal formation that supports the Soldier across full spectrum operations through a unified network architecture that is common across all Army echelons.

A-4. A TTSB also provides C2 to assigned and attached units while supervising the installation, operation and maintenance of communications nodes in the theater communications system excluding the division and corps systems.

A-5. It further provides real- and near real-time in-theater source information to combatant commanders and JTF commanders for the control, management, and dissemination of high volumes of data, to include air tasking orders, logistical, movement timetables, imagery, weather, etc. to deployed and dispersed forces in the theater.

A-6. Signal leaders (G-6/S-6) coordinate with supporting units (TTSBs) for inclusions in their network.

EXPEDITIONARY SIGNAL BATTALIONS

A-7. An ESB provides the Army and joint forces with an agile, expeditionary-capable signal formation that supports the Soldier across full spectrum operations through a unified network architecture that is common across all Army echelons.

A-8. An ESB operates 24 hours a day in austere environment to provide voice, data, and other network services to commanders previously conducted by theater, corps and division signal organizations. ESBs provide pooled signal assets to augment organic division/corps network support capabilities and/or replace network support battle losses at all echelons.

A-9. Signal leaders (G-6/S-6) coordinate with supporting units (ESBs) for inclusions in their network.

ADMINISTRATIVE AND LOGISTICS NETWORKS

A-10. Units establish A&L nets as required. Figure A-1 is an example of a typical division C2 network and Figure A-2 is an example of a brigade A&L FM network. All echelons, from battalion through division, have a support network to separate A&L from operational information. This prevents support information from overwhelming the C2 and O&I networks during operations.

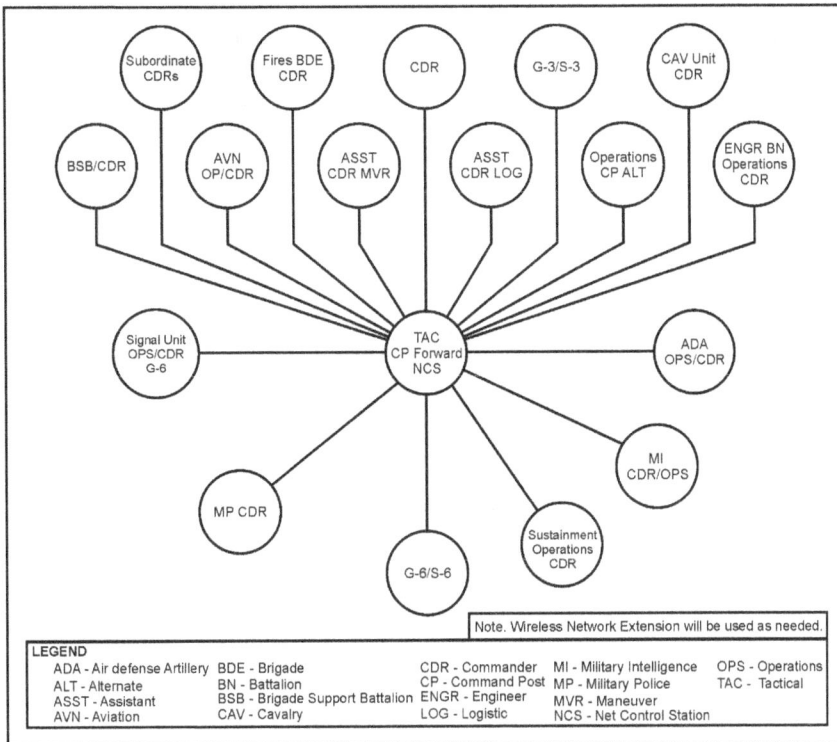

Figure A-1. Example of a division C2 FM network

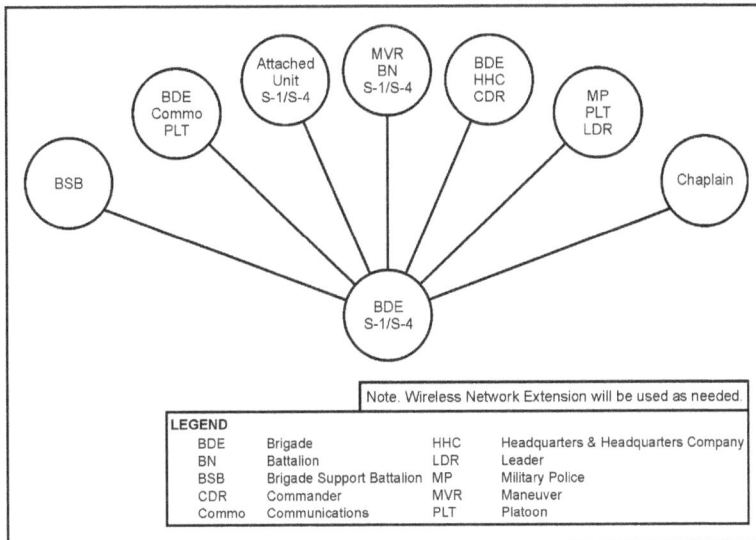

Figure A-2. Example of a brigade A&L FM network

OPERATIONS AND INTELLIGENCE NETWORKS

A-11. O&I nets are usually combined and established at brigade and battalion levels. Figure A-3 is an example of a division intelligence network. The information passed over these nets is continuous, and requires a separate net to prevent overloading the C2 net. The local situation determines whether other subscribers are added or deleted.

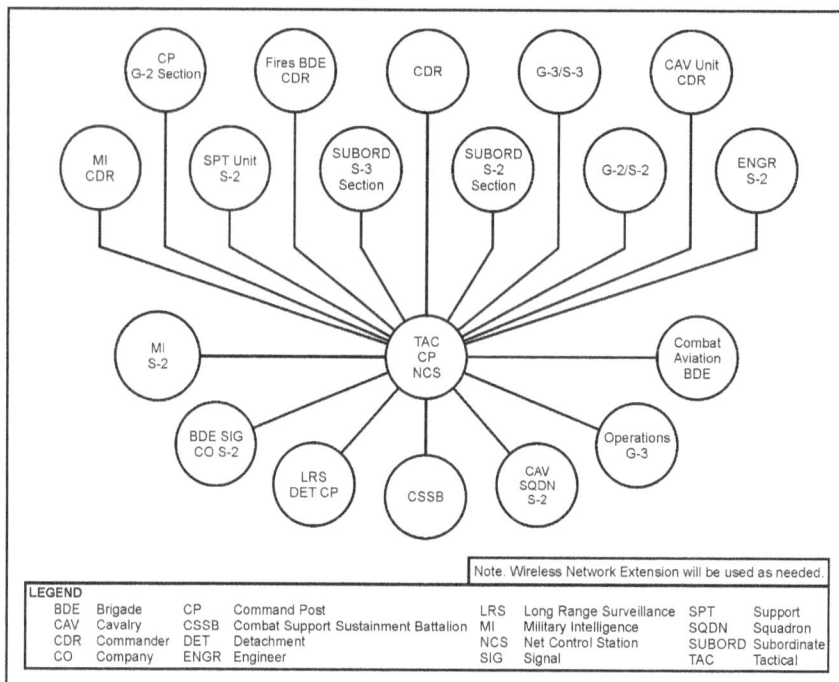

Figure A-3. Example of a division intelligence network

OTHER RELATED NETWORKS

A-12. Commanders may direct the G-6/S-6 to establish a variety of unit-specific networks dependent upon the commander's intent and METT-TC.

A-13. Wireless network extension operations extend the C2 network to ensure the availability of C2 at the critical moment during operations. In most cases, this network is established with the next higher headquarters.

HIGH FREQUENCY AND DATA NETWORKS

A-14. Data networks extend the tactical Internet to platforms that are not EPLRS equipped. Combat aviation brigades and air cavalry units use HF nets to provide long-range, non-LOS communications. Figure A-4 shows a typical cavalry unit HF net. Cavalry squadrons and troops use the low power HF for their C2 networks when distance is not an issue; the same is true of both divisional and regimental cavalry.

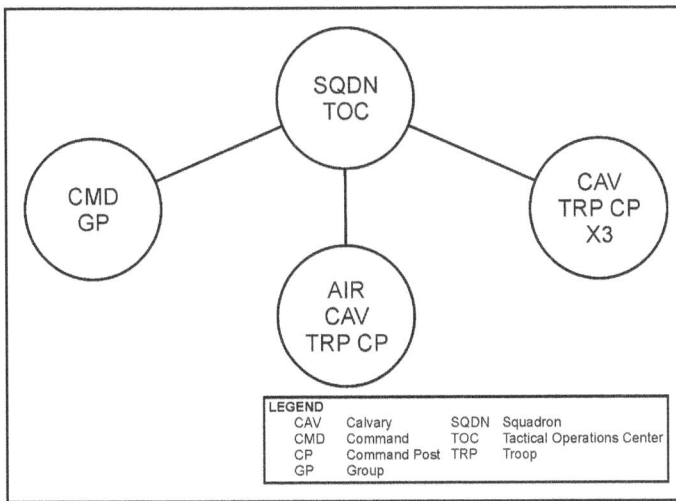

Figure A-4. Example of a cavalry unit HF network

Brigade Combat Team

A-15. The traditional HF nets are C2, A&L, O&I, fires, and other specialty uses such as reconnaissance. These nets were once limited due to the small number of HF radios available. Now, a brigade typically has between 70–80 HF radios and can establish nets down to the company and lower levels when the situation warrants it.

Medical Network

A-16. Medical units need dedicated, long-range, reliable communications systems that can be user-operated. Communications distances will be substantial between major medical support bases and forward aid stations. ALE tuning (Harris 5000 series radios) and other simplified operating features make HF ideal for units with a limited number of signal personnel. Figures A-5 and A-6 are examples of a medical unit HF networks for corps and division.

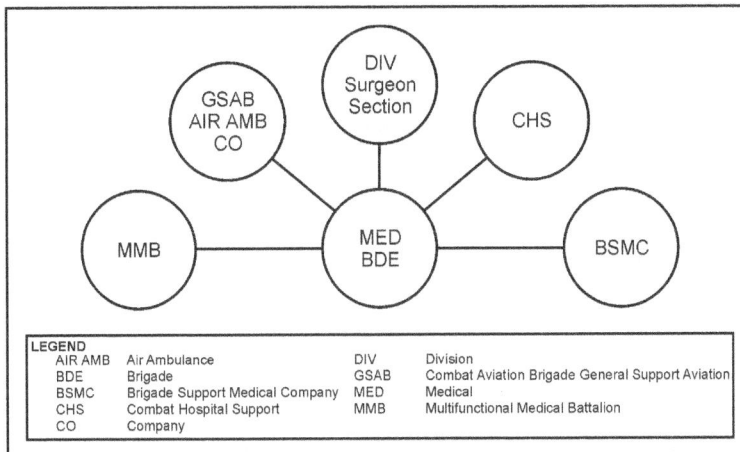

Figure A-5. Example of a division corps medical operations network HF-SSB

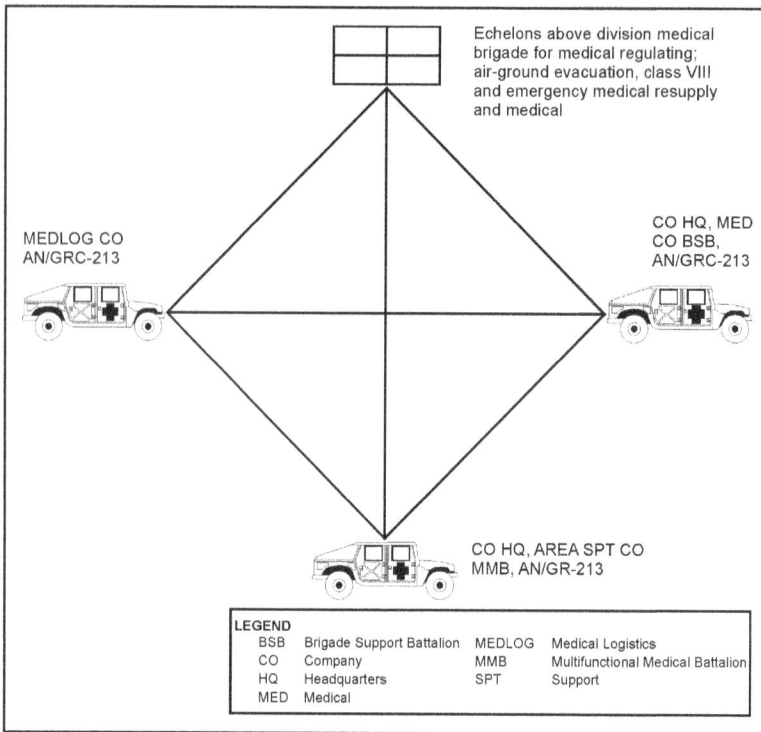

Figure A-6. Example of a medical operations network in a division HF-SSB

FIRE DIRECTION NETWORK

A-17. The fire direction network is the highest priority net in field artillery firing units. This network is used for exchange of technical and/or firing data. (Wireless network extension teams are also used to support these nets when needed.) Refer to TC 2-33.4 or FM 3-09.21 for more information on fire direction networks.

SURVEILLANCE NETWORK

A-18. The surveillance network passes along reports dealing with adversary movement and massing. The battalion battlefield information control center sets up this net to coordinate and control the ground surveillance radar and unattended ground sensor teams. The information from this net is vital to commanders and is given high priority for activation. Refer to FM 2-0 or FM 2-33.4 for more information on surveillance nets.

SUSTAINMENT AREA BATTLE COMMAND NETWORK

A-19. Sustainment area operations ensure freedom of maneuver. They consist of actions taken by Army units and host nation units (singularly or in a combined effort) to secure the force, or to neutralize or defeat adversary operations in the sustainment area. The sustainment area battle command FM net is a form of the C2 network. This network consists of many units that are collocated in the division sustainment area. Figure A-7 is an example of a division sustainment area FM network. Members of the sustainment area battle command network also depend on themselves to form the base cluster defense.

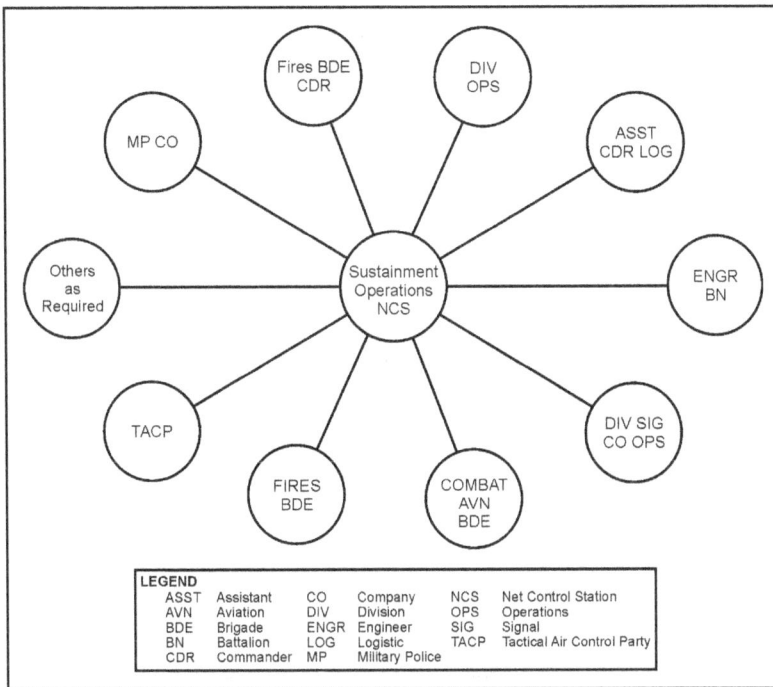

Figure A-7. Example of a division sustainment area FM network

HIGH FREQUENCY NETWORKS

A-20. The IHFRs (AN/PRC-104, AN/GRC-213, and AN/GRC-193) are being replaced by HF radios with ALE such as the AN/PRC-150 I. The HF nets shown are generic networks. Specific networks established, and subscribers to those networks, depend on command guidance and mission requirements.

A-21. HF networks are similar to the VHF FM networks in function and establishment. Many HF networks are a backup or supplement to their VHF FM counterparts. HF networks are established when unit dispersal exceeds the planning range for VHF FM systems. Figure A-8 is an example of a HF C2 network at division level. Note the similarity with the VHF FM C2 network. Commanders routinely establish a HF C2 network as a secondary means of controlling operations.

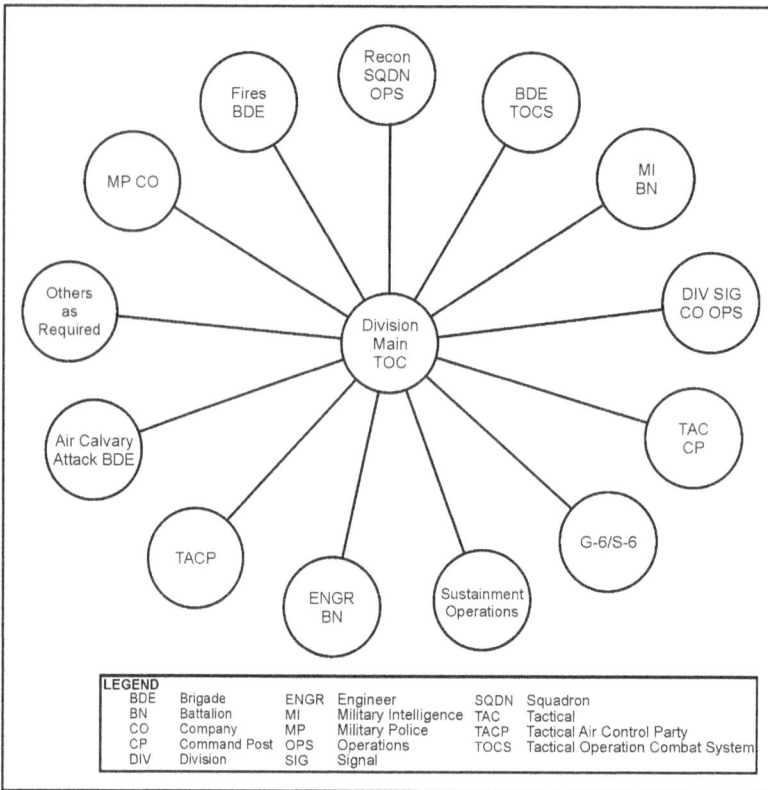

Figure A-8. Example of a division HF C2 network

A-22. Logistics units may use HF radios for C2 and internal coordination due to the communications distances from the division support area to the brigade support area. HF nets are a backup to FM networks, when the tactical spread of the division extends the lines of communications. The support units within the corps establish similar networks, or monitor the division networks to ensure push forward support.

Appendix B

Single-Channel Radio Communications Principles

SC radio communications equipment is used to transmit and receive voice, data, or telegraphic/voice code. This appendix addresses a radio sets basic components, characteristics and properties of radio waves, wave modulation, and site considerations for SC radios.

RADIO SET BASIC COMPONENTS

B-1. A radio set consists of a transmitter and receiver. Other items necessary for operation include a source of electrical power and an antenna for both radiation and reception of radio waves.

B-2. The transmitter contains an oscillator that generates RF energy in the form of alternating current. A transmission line, or cable, feeds the RF to the antenna. The antenna converts the alternating current into electromagnetic energy that is radiated into space; a keying device is used to control the transmission.

B-3. Normally, in SC radio operations, the receiver uses the same antenna as the transmitter to receive electromagnetic energy. The antenna converts the received electromagnetic energy into RF alternating current. The RF is fed to the receiver by a transmission line or cable. In the receiver, the RF is converted to audio frequencies. The audio frequencies are then changed into sound waves by a headset or loudspeaker.

B-4. Communications are possible when two radio sets operate on the same frequency, with the same type of modulation, and are within operating range.

RADIO TRANSMITTER

B-5. The simplest radio transmitter consists of a power supply and an oscillator. The power supply can be batteries, a generator, an alternating current power source with a rectifier and a filter, or a direct current rotating power source. The oscillator, which generates RF energy, must contain a circuit to tune the transmitter to the desired operating frequency. The transmitter must also have a device for controlling the emission of the RF signal. The simplest device is a telegraph key, a type of switch for controlling the flow of electric current. As the key is operated, the oscillator is turned on and off for varying lengths of time. The varying pulses of RF energy produced correspond to dots and dashes. This is a CW operation, and is used when transmitting international Morse code.

B-6. A CW radio transmitter is used to generate RF energy, which is radiated into space. The transmitter may contain only a simple oscillator stage. Usually, the output of the oscillator is applied to a buffer stage to increase oscillator stability, and to a PA that produces greater output. A telegraph key may be used to control the energy waves produced by the transmitter. When the key is closed, the transmitter produces its maximum output; when the key is opened, no output is produced.

B-7. By adding a modulator and a microphone, a radiotelephone transmitter can transmit messages by voice. When the modulating signal causes the amplitude of the radio wave to change, the radio is an AM set. When the modulating signal varies the frequency of the radio wave, the radio is an FM set.

Transmitter Characteristics

B-8. The reliability of radio communications depends on the characteristics of the transmitted signal. The transmitter, and its associated antenna, forms the initial step in the transfer of energy to a distant receiver.

B-9. Ground-wave transmission is used for most field radio communications. The range of the ground wave becomes correspondingly shorter as the operating frequency of the transmitter is increased through the applicable portions of the medium frequency (MF) band (300–3000 kHz) to the HF band (3.0–30 MHz). When the transmitter is operating at frequencies above 30 MHz, its range is generally limited to slightly more than LOS. For circuits using sky wave propagation, the frequency selected depends on the geographic area, season, and time of day.

Note. Frequency selection is the responsibility of the frequency manager not the RTO.

B-10. For maximum transfer of energy, the radiating antenna must be the proper length for the operating frequency. The local terrain determines, in part, the radiation pattern, and therefore affects the directivity of the antenna and the possible range of the set in the desired direction. When possible, several variations in the physical position of the antenna should be tried to determine the best operating position for radiating the greatest amount of energy in the desired direction.

B-11. The range of a transmitter is proportional to the power radiated by its antenna. An increase in the power output of the transmitter results in some increase in range. Under normal operating conditions, the transmitter should feed only enough power into the radiating antenna to establish reliable communications with the receiving station. Transmission of a signal more powerful than required is a breach of signal security, because adversary DF stations may instantly and more easily fix the location of the transmitter. Also, the signal can interfere with friendly stations operating on the same frequency.

RADIO RECEIVER

B-12. A radio receiver can receive modulated RF signals that carry speech, music, or other audio energy. It can also receive CW signals that are bursts of RF energy conveying messages by means of coded (dot/dash) signals.

B-13. The process of recovering intelligence from an RF signal is called detection; the circuit in which it occurs is called a detector. The detector recovers the intelligence from the carrier and makes it available for direct use, or for further amplification. In an FM receiver, the detector is usually called a discriminator.

B-14. An RF signal rapidly diminishes in strength after it leaves the transmitting antenna. Many RF signals of various frequencies are crowded into the RF spectrum. An RF amplifier selects and amplifies the desired signal; it contains integrated circuits or microprocessors to amplify the signal to a usable level. The RF amplifier is included in the receiver to sharpen the selectivity, and to increase the sensitivity. The RF amplifier normally uses tunable circuits to select the desired signal.

B-15. The signal level of the output of a detector, with or without an RF amplifier, is generally very low. One or more audio frequency amplifiers are used in the receiver, to build up the signal output to a useful level to operate headphones, a loudspeaker, or data devices.

Receiver Characteristics

B-16. When the transmitted signal reaches the receiver location, it arrives at a much lower power level than when it left the transmitter. The receiver must efficiently process this relatively weak signal to provide maximum reliability of communications.

B-17. Sensitivity describes how well a receiver responds to a weak signal at a given frequency. A receiver with high sensitivity is able to accept a very weak signal, and amplify and process it to provide a usable output. The principal factor that limits or lowers the sensitivity of a receiver is the noise generated by its own internal circuits.

B-18. Selectivity describes how well a receiver is able to differentiate between a desired frequency and undesired frequencies.

B-19. In field radio communications, the type, location, and electrical characteristics of the receiving antenna are not as important as they are for the transmitting antenna. The receiving antenna must be of sufficient length, be properly coupled to the input of the receiver circuit, and (except in some cases for HF sky wave propagation) must have the same polarization as the transmitting antenna.

RADIO WAVES

B-20. Radio waves travel near the surface of the earth, and radiate skyward at various angles to the earth's surface. These electromagnetic waves travel through space at the speed of light, approximately 300,000 km (186,000 miles) per second. Figure B-1 shows the wave radiation from a vertical antenna.

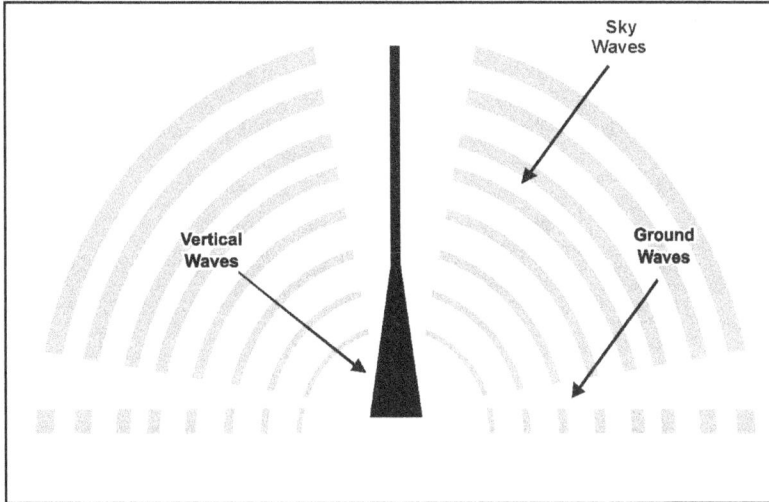

Figure B-1. Radiation of radio waves from a vertical antenna

WAVELENGTH

B-21. The wavelength is defined as the distance between the crest of one wave to the crest of the next wave; it is the length (always measured in meters) of one complete cycle of the waveform. Figure B-2 shows the wavelength of a radio wave.

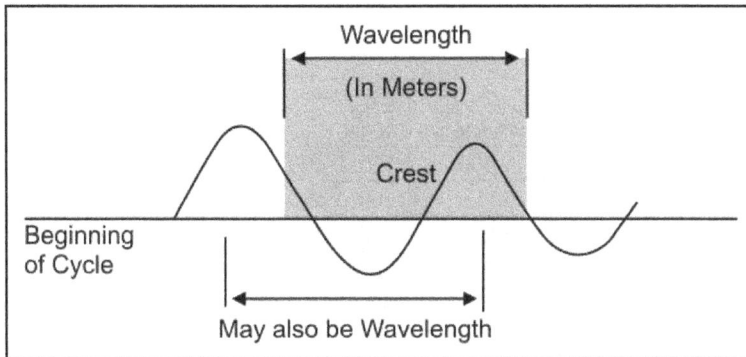

Figure B-2. Wavelength of a radio wave

FREQUENCY

B-22. The frequency of a radio wave is the same as the number of complete cycles that occur in one second. The longer the time of one cycle, the longer the wavelength and the lower the frequency; frequency is measured and stated in Hz. One cycle per second is stated as 1 Hz. Because the frequency of a radio wave is very high, it is generally measured and stated in kHz (one thousand hertz) or MHz (one million hertz) per second. Sometimes frequencies are expressed in GHz (one billion hertz) per second.

Frequency Calculation

B-23. For practical purposes, the velocity of a radio wave is considered constant, regardless of the frequency or the amplitude of the transmitted wave. Therefore, to find the frequency when the free-space wavelength is known, divide the velocity by the wavelength, for example—

- **Frequency (Hz)** = 300,000,000 (meters per second) wavelength in meters.
- **Wavelength (meters)** = 300,000,000 (meters per second) frequency in Hz.

Frequency Bands

B-24. Within the RF spectrum, radio frequencies are divided into groups, or bands, of frequencies. Table B-1 lists the frequency band coverage. Most tactical radio sets operate within a 2–400 MHz range within the frequency spectrum.

Table B-1. Frequency band chart

Band	Frequency
Very low frequency	3–30 kHz
Low frequency	30–300 kHz
MF	.3–3.0 MHz
HF	3.0–30 MHz
VHF	30–300 MHz
UHF	300–3,000 GHz
Super high frequency	3,000–30,000 GHz
EHF	30,000–300,000 GHz

B-25. Table B-2 lists certain characteristics of each frequency band. The ranges and power requirements shown are for normal operating conditions (proper site selection and antenna orientation, and correct operating procedures). The ranges will change according to the condition of the propagation medium and the transmitter output power.

Table B-2. Frequency band characteristics

Band	Range				Power Required (Kilowatt [kW])
	Ground Wave		Sky Wave		
	Miles	Kilometers	Miles	Kilometers	
Low frequency	0–1,000	0–1,609	500–8,000	805–12,872	Above 50
MF	0–100	0–161	100–1,500	161–2,415	.5–50
HF	0–50	0–83	100–8,000	161–12,872	.5–5
VHF	0–30	0–48	50–150	80.5–241	.5 or Less
UHF	0–50	0–83	unlimited (refer to paragraph B-30)		.5 or Less

B-26. The frequency of the radio wave affects its propagation characteristics. At low frequencies (.03–.3 MHz), the ground wave is very useful for communications over great distances. The ground wave signals are quite stable and show little seasonal variation.

B-27. In the MF band (.3–3.0 MHz) the range of the ground wave varies from about 24 km (15 miles) at 3 MHz to about 640 km (400 miles) at the lowest frequencies of this band. Sky wave reception is possible during the day or night at any of the lower frequencies in this band. At night, the sky wave is receivable at distances up to 12,870 km (8,000 miles). Major uses of the MF band include medium distance communications, radio navigation, and AM broadcasting.

B-28. In the HF band (3.0–30 MHz), the range of the ground wave decreases as frequency increases, and the sky waves are greatly influenced by ionospheric considerations. HF is widely used for long distance communications, short-wave broadcasting, and over-the-horizon radar; HF is also used to supplement tactical communications when LOS communication is not possible or feasible.

B-29. In the VHF band (30–300 MHz), there is no usable ground wave and only slight refraction of sky waves by the ionosphere at the lower frequencies. The direct wave (LOS) provides communications if the transmitting and receiving antennas are elevated high enough above the surface of the Earth.

B-30. In the UHF band (300–3,000 GHz), the direct wave must be used for all transmissions (15–100 miles). Communications are limited to a short distance beyond the horizon. Lack of static and fading in these bands makes LOS reception satisfactory. Antennas that are highly directional can be used to concentrate the beam of RF energy, thus increasing the signal intensity. UHF satellite transmissions can cover thousands of miles, depending on altitude, power, and antenna configuration.

PROPAGATION

B-31. Ground waves and sky waves are the two principal paths by which radio waves travel from a transmitter to a receiver. Figure B-3 is an example of the principal paths of radio waves. Ground waves travel directly from the transmitter to the receiver; sky waves travel up to the ionosphere and are refracted (bent downward) back to the earth. Short distance, UHF, and upper VHF transmissions are made by ground waves; long distance transmission is principally by sky waves. SC radio sets can use either ground wave or sky wave propagation for communications.

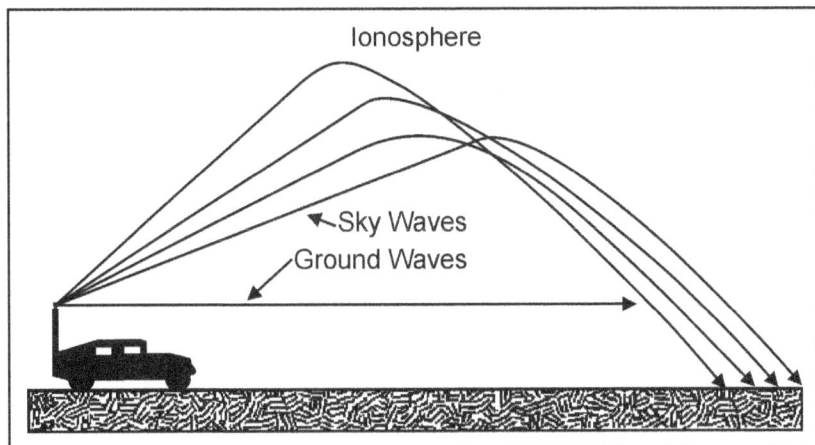

Figure B-3. Principal paths of radio waves

GROUND WAVE PROPAGATION

B-32. Radio communications that use ground wave propagation do not use or depend on waves that are refracted from the ionosphere (sky waves). Ground wave propagation is affected by the electrical characteristics of the earth and the amount of diffraction (bending) of the waves along the curvature of the earth. The strength of the ground wave at the receiver depends on the power output and frequency of the transmitter, the shape and conductivity of the earth along the transmission path, and the local weather conditions. Figure B-4 shows possible routes for ground waves.

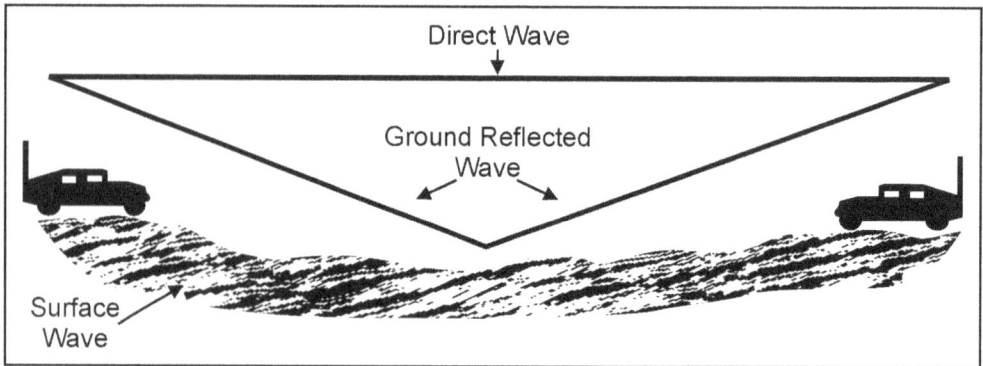

Figure B-4. Possible routes for ground waves

Direct Wave

B-33. The direct wave travels directly from the transmitting antenna to the receiving antenna. The direct part of the wave is limited to the LOS distance between the transmitting and receiving antennas, and the small distance added by atmospheric refraction and diffraction of the wave around the curvature of the Earth. Increasing the height of the transmitting or receiving antenna, or both, can extend this distance.

Ground-Reflected Wave

B-34. The ground wave reaches the receiving antenna after being reflected from the surface of the earth. Cancellation of the radio signal can occur when the ground reflected component and the direct wave component arrive at the receiving antenna at the same time, and are 180 degrees out of phase with each other.

Surface Wave

B-35. The surface wave follows the Earth's curvature and is affected by the Earth's conductivity and dielectric constant.

FREQUENCY CHARACTERISTICS OF GROUND WAVES

B-36. Various frequencies determine which wave component will prevail along any given signal path. For example, when the Earth's conductivity is high and the frequency of a radiated signal is low, the surface wave is the predominant component. For frequencies below 10 MHz, the surface wave is sometimes the predominant component. However, above 10 MHz, the losses that are sustained by the surface wave component are so great that the other components (direct and sky wave) become predominant.

B-37. At frequencies of 30–300 kHz, ground losses are very small, so the surface wave component follows the Earth's curvature. It can be used for long-distance communications provided the RTO has enough power from the transmitter. The frequencies 300 kHz–3 MHz are used for long distance communications over sea water and for medium-distance communications over land.

B-38. At HF, 3–30 MHz, the ground's conductivity is extremely important, especially above 10 MHz where the dielectric constant or conductivity of the Earth's surface determines how much signal absorption occurs. In general, the signal is strongest at the lower frequencies when the surface over which it travels has a high dielectric constant and conductivity.

Earth's Surface Conductivity

B-39. The dielectric constant or Earth's surface conductivity determines how much of the surface wave signal energy will be absorbed or lost. Although the Earth's surface conductivity as a whole is generally poor, Table B-3 shows a comparison of the conductivity of varying surface conditions.

Table B-3. Surface conductivity

Surface Type	Relative Conductivity
Large body of fresh water	Very good
Ocean or sea water	Good
Flat or hilly loamy soil	Fair
Rocky terrain	Poor
Desert	Poor
Jungle	Very poor

SKY WAVE PROPAGATION

B-40. Radio communications that use sky wave propagation depend on the ionosphere to provide the signal path between the transmitting and receiving antennas. The ionosphere has four distinct layers. These layers are labeled D, E, F1, and F2, in the order of increasing heights and decreasing molecular densities. During the day, when the rays of the sun are directed toward that portion of the atmosphere, all four layers may be present. During the night, the F1 and F2 layers seem to merge into a single F layer, while the D and E layers fade out. The actual number of layers, their height above the earth, and their relative intensity of ionization, varies constantly. Table B-4 provides a description of the ionosphere layers and Figure B-5 shows the average layer distribution of the ionosphere.

Table B-4. Ionosphere layers

Region	Description
D Region	Exists only during daylight hours and has little effect in bending the paths of HF radio waves. The main effect of the D region is to attenuate HF waves when the transmission path is in sunlit regions.
E Region	Is used during the day for HF radio transmission over intermediate distances (less than 2,400 km [1,500 miles]). At night, the intensity of the E region decreases and it becomes useless for radio transmission.
F Region	Exists at heights up to 380 km (240 miles) above the earth and is ionized all the time. It has two well-defined layers (F1 and F2) during the day and one layer (F) during the night. At night, the F region remains at a height of about 260 km (170 miles) and is useful for long-range radio communications (over 2,400 km [1,500 miles]). The F2 layer is the most useful of all layers for long-range radio communications, although its degree of ionization varies appreciably from day to day.

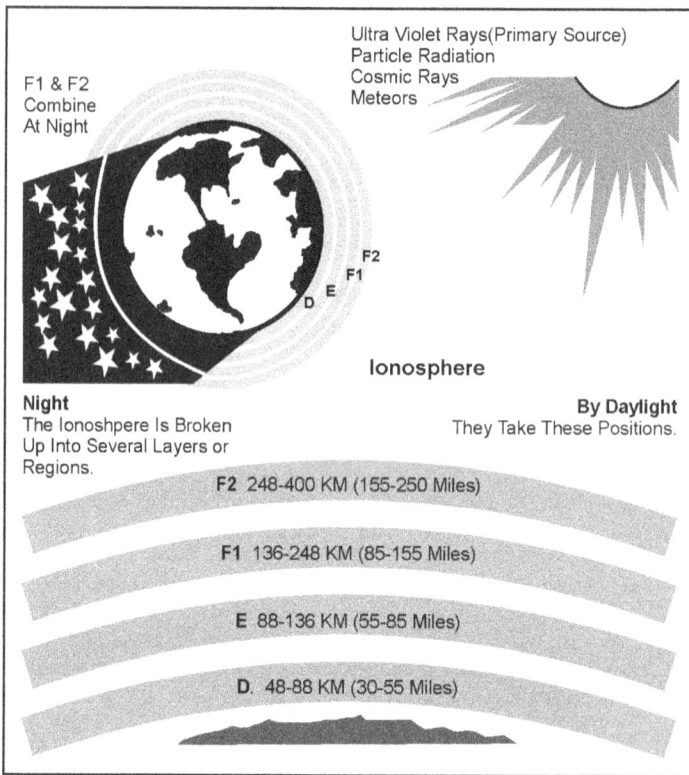

Figure B-5. Average layer distribution of the ionosphere

B-41. The movement of the earth around the sun, and changes in the sun's activity, contribute to ionospheric variations. These variations are regular, and therefore predictable; and irregular, which occur from abnormal behavior of the sun. Table B-5 lists the regular variations of the ionosphere.

Table B-5. Regular variations of the ionosphere

Variation	Description
Daily	Caused by the rotation of the earth.
Seasonal	Caused by the north and south progression of the sun.
27-day	Caused by the rotation of the sun on its axis.
11-year	Caused by the sunspot activity cycle going from maximum through minimum back to maximum levels of intensity.

B-42. In planning a communications system, the status of the four regular variations must be anticipated. Irregular variations must also be considered since they have a degrading effect (at times blanking out communications), which currently cannot be controlled or compensated for. Table B-6 lists some irregular variations of the ionosphere.

Table B-6. Irregular variations of the ionosphere

Variation	Description
Sporadic E	When excessively ionized, the E layer often blanks out the reflections back from the higher layers. It can also cause unexpected propagation of signals hundreds of miles beyond the normal range. This effect can occur at any time.
Sudden Ionospheric Disturbance	Coincides with a bright solar eruption, and causes abnormal ionization of the D layer. This effect causes total absorption of all frequencies above approximately 1 MHz. It can occur without warning during daylight hours, and can last from a few minutes to several hours. When it occurs, receivers seem to go dead.
Ionospheric Storms	During these storms, sky wave reception above approximately 1.5 MHz shows low intensity, and is subject to a type of rapid blasting and fading called flutter fading. May last from several hours to several days, and usually extend over the entire earth.

B-43. Sunspots generate bursts of radiation that cause high levels of ionization. The more sunspots, the greater the ionization. During periods of low sunspot activity, frequencies above 20 MHz tend to be unusable because the E and F layers are too weakly ionized to reflect signal back to Earth. At the peak of the sunspot cycle, however, it is unusual to have worldwide propagation on frequencies above 30 MHz.

B-44. Primarily, the ionization density of each layer determines the range of long distance radio transmissions; the higher the frequency, the greater the ionization density required to reflect radio waves back to earth. The upper (E and F) regions reflect the higher frequencies, because they are the most highly ionized. The D region, which is the least ionized, does not reflect frequencies above approximately 500 kHz. Thus, at any given time and for each ionized region, there is an upper frequency limit at which radio waves sent vertically upward are reflected back to earth. This limit is called the critical frequency.

B-45. Radio waves directed vertically at frequencies higher than the critical frequency pass through the ionized layer out into space. All radio waves that are directed vertically into the ionosphere at frequencies lower than the critical frequency are reflected back to earth.

B-46. Generally, radio waves used in communications are directed toward the ionosphere at some oblique angle, called the angle of incidence. Radio waves at frequencies above the critical frequency will be reflected back to earth if transmitted at angles of incidence smaller than a certain angle, called the critical angle. At the critical angle, and at all angles larger than the critical angle, the radio waves will pass through the ionosphere if the frequency is higher than the critical frequency.

TRANSMISSION PATHS

B-47. Sky wave propagation refers to those types of radio transmissions that depend on the ionosphere to provide signal paths between transmitters and receivers. Figure B-6 shows the sky wave transmission paths. The distance from the transmitting antenna to the place where the sky waves first return to earth is called the skip distance. The skip distance depends upon the angle of incidence, the operating frequency, and the height and density of the ionosphere.

B-48. The antenna height, in relation to the operating frequency, affects the angles at which transmitted radio waves strike and penetrate the ionosphere and then return to Earth. This angle of incidence can be controlled to obtain the desired area of coverage; lowering the antenna height will increase the angle of transmission. This provides broad and even signal patterns in an area the size of a typical corps. The use of near-vertical transmission paths is known as NVIS. Raising the antenna height will lower the angle of incidence.

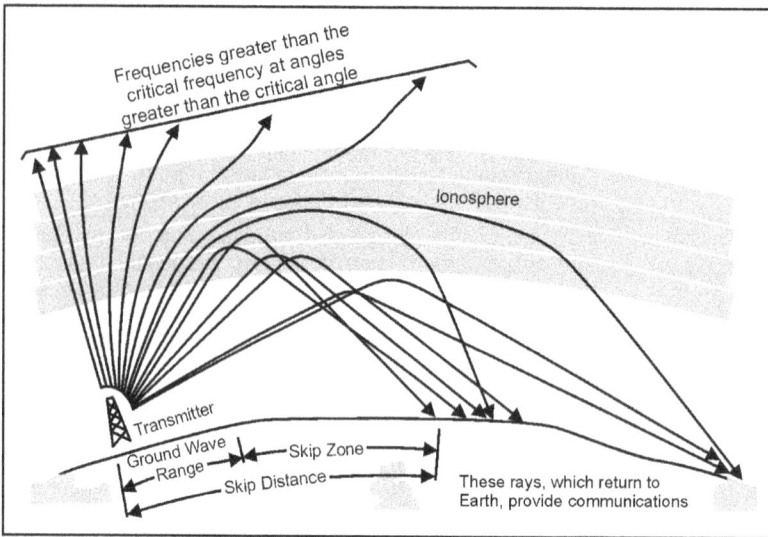

Figure B-6. Sky wave transmission paths

B-49. Lowering the angle of incidence can produce a skip zone in which no usable signal can be received. This area is bounded by the outer edge of usable ground wave propagation and the point nearest the antenna at which the sky wave returns to earth. In corps area communications situations, the skip zone is not a desirable condition. However, low angles of incidence make long distance communications possible.

B-50. When a transmitted wave is reflected back to the surface of the earth, the earth absorbs part of its energy. The remainder of its energy is reflected back into the ionosphere and reflected back to earth again. This means of transmission (by alternately reflecting the radio wave between the ionosphere and the earth) is called hops. Hops enable radio waves to be received at great distances from the point of origin. Figure B-7 is an example of sky wave transmission hop paths.

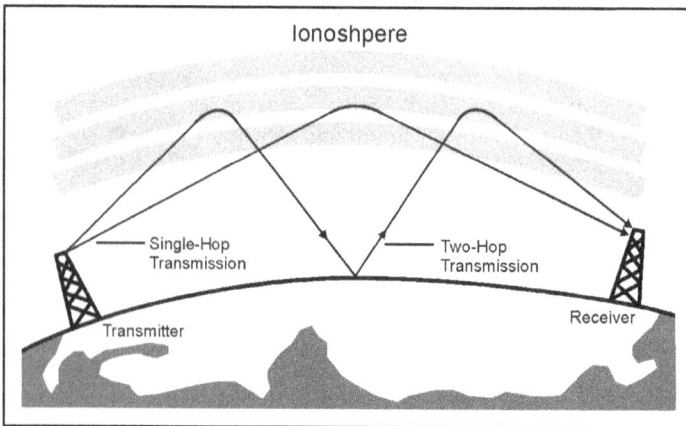

Figure B-7. Sky wave transmission hop paths

Fading

B-51. Fading is the periodic increase and decrease of received signal strength. Fading occurs when a radio signal is received over a long distance path in the HF range. The precise origin of this fading is seldom understood. There is little common knowledge of what precautions to take to reduce or eliminate fading's troublesome effects. Fading associated with sky wave paths is the greatest detriment to reliable communications. Too often, those responsible for communications circuits rely on raising the transmitter power or increasing antenna gain to overcome fading. Unfortunately, such actions often do not work and seldom improve reliability. Only when the signal level fades down below the back ground noise level for an appreciable fraction of time will increased transmitter power or antenna gain yield an overall circuit improvement. Choosing the correct frequency and using transmitting and receiving equipment intelligently ensure a strong and reliable receiving signal, even when low power is used.

Maximum Usable Frequency and Lowest Usable Frequency

B-52. The maximum usable frequency (MUF) is the maximum frequency at which a radio wave will return to earth at a given distance, when using a given ionized layer and a transmitting antenna with a fixed angle of radiation. It is the monthly median of the daily highest frequency that is predicted for sky wave transmission over a particular path at a particular hour of the day. The MUF is always higher than the critical frequency because the angle of incidence is less than 90 degrees.

B-53. If the distance between the transmitter and the receiver is increased, the MUF will also increase. Radio waves lose some of their energy through absorption by both the D region, and a portion of the E region of the ionosphere, on certain transmission frequencies. The total absorption is less, and communications more satisfactory, as higher frequencies are used up to the level of the MUF.

B-54. The absorption rate is greatest for frequencies ranging from approximately 500 kHz–2 MHz during the day. During the night, the absorption rate decreases for all frequencies. As the frequency of transmission over any skywave path is increased from low to high, a frequency will be reached at which the received signal overrides the level of atmospheric and other radio noise interference. This is called the lowest usable frequency, because frequencies lower than these are too weak for useful communications. It should be noted that the lowest usable frequency also depends on the power output of the transmitter, and the transmission distance. When the lowest usable frequency is greater than the MUF, no sky wave transmission is possible. The frequency manager uses SPECTRUM XXI to identify optimum frequency groupings.

Other Factors Affecting Propagation

B-55. In VHF and UHF ranges, extending from 30–300 MHz and beyond, the presence of object (buildings or towers for example) may produce strong reflections that arrive at the receiving antenna in such a way that they cancel the signal from the desired propagation path and render communications impossible.

B-56. Receiver locations that avoid the proximity of an airfield should be chosen due to possible adverse interference from signals bouncing off of the aircrafts. Avoid locating transmitters and receivers where an airfield is at or near midpoint of the propagation path of frequencies above 20 MHz.

B-57. Many other factors may affect the propagation of a radio wave. Hills, mountains, buildings, water towers, tall fences, and even other antenna can have a marked affect on the condition and reliability of a given propagation path. Conductivity of the local ground or body of water can greatly alter the strength of the transmitted or received signal. Energy radiation from the Sun's surface also greatly affects conditions within the ionosphere and alters the characteristics of long-distance propagation at 2–30 MHz.

Path Loss

B-58. Radio waves become weaker as they spread outwards from the transmitter. The ratio of the received power is called path loss. LOS paths at VHF and UHF require relatively little power since the total path loss at the radio horizon is only about 25 dB greater than the path loss over the same distance in free space (absence of ground). This additional loss results from some energy being reflected from the ground, canceling part of the direct wave energy. This is unavoidable in almost every practical case. The total path loss for an LOS path above average terrain varies with the following factors: total path loss between transmitting and receiving antenna terminals, frequency, distance, transmitting antenna gain, and receiving antenna gain.

Reflected Waves

B-59. Often, it is possible to communicate beyond the normal LOS distance by exploiting the reflection from a tall building, nearby mountain, or water tower. If the top portion of a structure or hill can be seen readily by both transmitting and receiving antennas, it may be possible to achieve practical communications by directing both antennas toward the point of maximum reflection. If the reflecting object is very large in terms of a wavelength, the path loss, including the reflection, can be very low.

B-60. If a structure or hill exists adjacent to an LOS path, reflected energy may either add to or subtract from the energy arriving from the direct path. If the reflected energy arrives at the receiving antenna with the same amplitude (strength) as the direct signal but has the opposite phase, both signals will cancel and communication will be impossible. However, if the same condition exists but both signals arrive in phase, they will add and double the signal strength. These two conditions represent destructive and constructive combinations of the reflected and direct waves.

B-61. Reflection from the ground at the common midpoint between the receiving and transmitting antennas may also arrive in a constructive or destructive manner. Generally, in the VHF and UHF range, the reflected wave is out of phase (destructive) with respect to the direct wave at vertical angles less than a few degrees above the horizon. However, since the ground is not a perfect conductor, the amplitude of the reflected wave seldom approaches that of the direct wave. Thus, even though the two arrive out of phase, complete cancellation does not occur. Some improvement may result from using vertical polarization rather than horizontal polarization over LOS paths because there tends to be less phase difference between direct and reflected waves. The difference is usually less than 10 dB, however, in favor of vertical polarization.

Diffraction

B-62. Unlike the ship passing beyond the visual horizon, a radio wave does not fade out completely when it reaches the radio horizon. A small amount of radio energy travels beyond the radio horizon by a process called diffraction. Diffraction also occurs when a light source is held near an opaque object, casting a shadow on a surface behind it. Near the edge of the shadow a narrow band can be seen which is neither completely light nor dark. The transition from total light to total darkness does not occur abruptly, but changes smoothly as the light is diffracted.

B-63. A radio wave passing over either the curved surface of the Earth or a mountain ridge behaves in much the same fashion as a light wave. For example, people living in a valley below a high, sharp, mountain ridge can often receive a TV station located many miles below on the other side. TV station are diffracted by the mountain ridge and bent downward in the direction of the town. It is emphasized, however, that the energy decays very rapidly as the angle of propagation departs from the straight LOS path. Typically, a diffracted signal may undergo a reduction of 30 to 40 dB by being bent only 5 ft (1.5 meters) by a mountain ridge. The actual amount of diffracted signal depends on the shape of the surface, the frequency, the diffraction angle, and many other factors. It is sufficient to say that there are times when the use of diffraction becomes practical as a means for communicating in the VHF and UHF over long distances.

Refraction

B-64. Refraction is the bending of a wave as it passes through air layers of different density (refractive index). In semitropical regions, a layer of air 5–100 meters (16.4–328 ft) (thick with distinctive characteristics may form close to the ground, usually the result of a temperature inversion. For example, on an unusually warm day after a rainy spell, the Sun may heat up the ground and create a layer of warm, moist air. After sunset, the air a few meters above the ground will cool very rapidly while the moisture in the air close to the ground serves as a blanket for the remaining heat. After a few hours, a sizable difference in temperature may exist between the air near the ground and the air at a height of 10–20 meters (32.8–65.6 ft) resulting in a marked difference in air pressure. Thus, the air near the ground is considerably denser than the air higher up. This condition may exist over an area of several hundred square kilometers or over a long area of land near a seacoast. When such an air mass forms, it usually remains stable until dawn, when the ground begins to cool and the temperature inversion ends.

B-65. When a VHF or UHF radio wave is launched within such air mass, it may bend or become trapped (forced to follow the inversion layer). This layer then acts as a duct between the transmitting antenna and a distant receiving site. The effects of such ducting can be seen frequently during the year in certain locations where TV or VHF FM stations are received over paths of several hundred kilometers. The total path loss within such a duct is usually very low and may exceed the free space loss by only a few dBs.

B-66. It is also possible to communicate over long distances by means of tropospheric scatter. At altitudes of a few kilometers, the air mass has varying temperature, pressure, and moisture content. Small fluctuations in tropospheric characteristics at high altitude create blobs. Within a blob, the temperature, pressure, and humidity are different from the surrounding air. If the difference is large enough, it may modify the refractive index at VHF and UHF. A random distribution of these blobs exists at various altitudes at all times. If a high-power transmitter (greater than 1 kW) and high gain antenna (10 dB or more) are used, sufficient energy may be scattered from these blobs down to the receiver to make reliable communication possible over several hundred kilometers. Communication circuits employing this mode of propagation must use very sensitive receivers and some form of diversity to reduce the effects of the rapid and deep fading. Scatter propagation is usually limited to path distances of less than about 500 km (310.6 miles).

Noise

B-67. Noise consists of all undesired radio signals, manmade or natural. Noise masks and degrades useful information reception. The radio signal's strength is of little importance if the signal power is greater than the received noise power. This is why S/N ratio is the most important quantity in a receiving system. Increasing receiver amplification cannot improve the S/N ratio since both signal and noise will be amplified equal and S/N ratio will remain unchanged. Normally, receivers have more than enough amplification.

B-68. Natural noise has two principle sources: thunderstorms (atmospheric noise) and stars (galactic noise). Both sources generate sharp pulses of electromagnetic energy over all frequencies. The pulses propagate according to the same laws as manmade signals, and receiving systems must accept them along with the desired signal. Atmospheric noise is dominant from 0–5 MHz, and galactic noise is most important at higher frequencies. Low frequency transmitters must generate very strong signals to overcome noise. Strong signals and strong noise mean that the receiving antenna does not have to be large to collect a usable signal. A 1.5 meter (4.9 ft) tuned whip antenna will adequately deliver all of the signals that can be received at frequencies below 1 MHz.

B-69. Manmade noise is a product of urban civilization that appears wherever electric power is used. It is generated anywhere that there is an electric arc (automobile, power lines, motors or fluorescent lights). Each source is small, but there are so many that together they can completely hide a weak signal that would be above the natural noise in rural areas. Manmade noise is troublesome when the receiving antenna is near the source, but being near the source gives the noise waves characteristics that can be exploited. Waves near a source tend to be vertically polarized. A horizontally polarized receiving antenna will generally receive less noise than a vertically polarized antenna.

B-70. Manmade noise currents are induced by any conductors near the source, including the antenna, transmission line, and equipment cases. If the antenna and transmission line are balanced with respect to the ground, then the noise voltages will be balanced and cancel with respect to the receiver input terminals (zero voltage across terminals), and this noise will not be received. Near perfect balance is difficult to achieve, but any balance may help.

B-71. Other ways to avoid manmade noise are to locate the most troublesome sources and turn them off, or move the receiving system away from them. Moving at least one km (.6 miles) away from a busy street or highway will significantly reduce noise. Although broadband receiving antennas are convenient because they do not have to be tuned to each working frequency, sometimes a narrowband antenna can make the difference between communicating and not communicating. The HF band is now so crowded with users that interference and noise, not signal strength, are the main reasons for poor communications. A narrowband antenna will reject strong interfering signals near the desired frequency and help maintain good communications.

WAVE MODULATION

B-72. Both FM and AM transmitters produce RF carriers. The carrier is a wave of constant amplitude, frequency, and phase which can be modulated by changing its amplitude, frequency, or phase. Thus, the RF carrier carries intelligence by being modulated. Modulation is the process of superimposing intelligence (voice or coded signals) on the carrier. Figure B-8 shows different wave shapes.

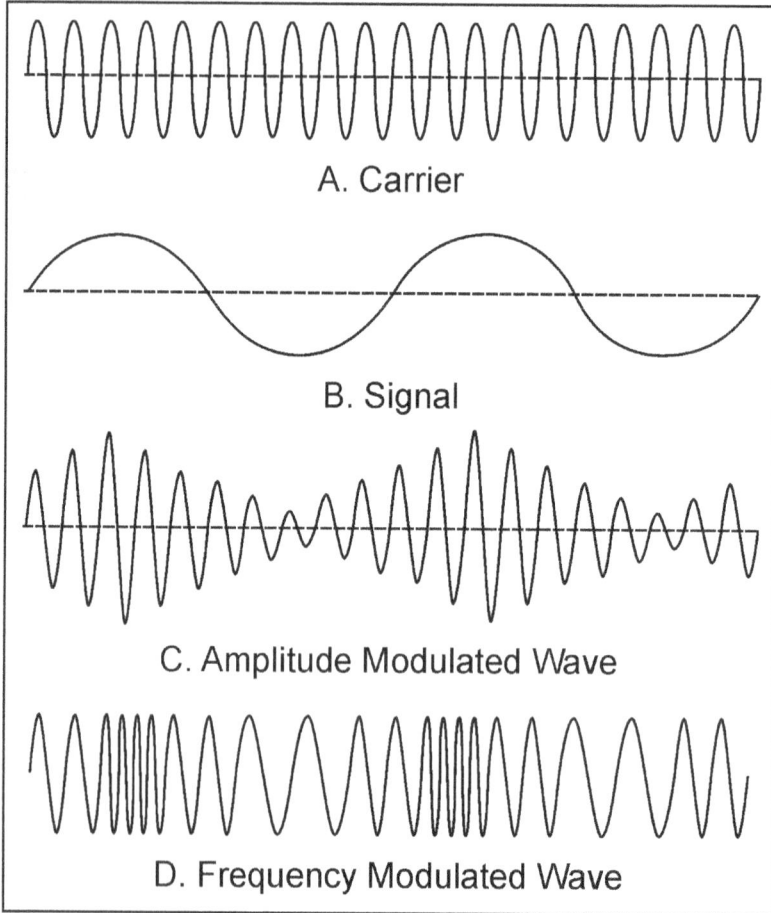

A. Carrier

B. Signal

C. Amplitude Modulated Wave

D. Frequency Modulated Wave

Figure B-8. Wave shapes

FREQUENCY MODULATION

B-73. FM is the process of varying the frequency (rather than the amplitude) of the carrier signal in accordance with the variations of the modulating signals. The amplitude or power of the FM carrier does not vary during modulation. The frequency of the carrier signal, when it is not modulated, is called the center, or rest, frequency. When a modulating signal is applied to the carrier, the carrier signal will move up and down in frequency away from the center, or rest, frequency.

B-74. The amplitude of the modulating signal determines how far away from the center frequency the carrier will move. This movement of the carrier is called deviation; how far the carrier moves is called the amount of deviation. During reception of the FM signal, the amount of deviation determines the loudness or volume of the signal.

B-75. The FM signal leaving the transmitting antenna is constant in amplitude, but varies in frequency according to the audio signal. As the signal travels to the receiving antenna, it picks up natural and manmade electrical noises that cause amplitude variations in the signal. All of these undesirable amplitude variations are amplified as the signal passes through successive stages of the receiver, until the signal reaches a part of the receiver called the limiter. The limiter is unique to FM receivers, as is the discriminator.

B-76. The limiter eliminates the amplitude variations in the signal, and then passes it on to the discriminator, which is sensitive to variations in the frequency of the RF wave. The resultant constant amplitude FM signal is then processed by the discriminator circuit, which changes the frequency variations into corresponding voltage amplitude variations. These voltage variations reproduce the original modulating signal in a headset, loudspeaker, or teletypewriter. Radiotelephone transmitters operating in the VHF and higher frequency bands generally use FM.

AMPLITUDE MODULATION

B-77. AM is the variation of the RF power output of a transmitter at an audio rate. Stated differently, the RF energy increases and decreases in power, according to the audio frequencies superimposed on the carrier signal.

B-78. When audio frequency signals are superimposed on the RF carrier signal, additional RF signals are generated. These additional frequencies are equal to the sum and the difference of the audio frequency and RF used. For example, assume a 500 kHz carrier is modulated by a one kHz audio tone. Two new frequencies are developed, one at 501 kHz (the sum of 500 kHz and one kHz) and the other at 499 kHz (the difference between 500 kHz and 1 kHz). If a complex audio signal is used instead of a single tone, two new frequencies will be created for each of the audio frequencies involved. New frequencies resulting from superimposing an audio frequency signal on a RF signal are called sidebands.

B-79. When the RF carrier is modulated by complex tones, such as speech, each separate frequency component of the modulating signal produces its own upper and lower sideband frequencies. The upper sideband contains the sum of the RF and audio frequency signals, and the lower sideband contains the difference between the RF and audio frequency signals. Figure B-9 shows an AM system.

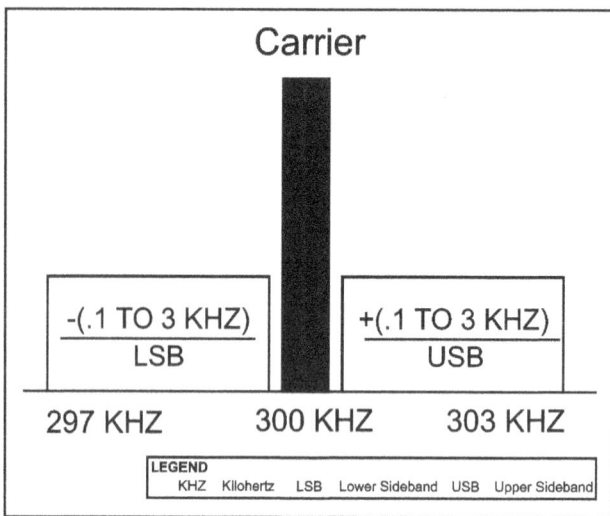

Figure B-9. AM system

B-80. The space occupied by a carrier and its associated sidebands in the RF spectrum is called a channel. In AM, the width of the channel (bandwidth) is equal to twice the highest modulating frequency. For example, if a 5,000 kHz (5 MHz) carrier is modulated by a band of frequencies ranging from 200–5,000 cycles (.2–5 kHz); the upper sideband extends from 5000.2–5005 kHz. The lower sideband extends from 4,999.8–4,995 kHz. Thus, the bandwidth is the difference between 5,005 Hz–4,995 kHz, a total of 10 kHz.

B-81. Radiotelephone transmitters operating in the medium and HF bands generally use AM; the intelligence of an AM signal exists solely in the sidebands.

SINGLE SIDE BAND

B-82. Each sideband contains all the intelligence needed for communications. Although both sidebands are generated within the modulation circuitry of the SSB radio set, the carrier and one sideband are removed before any signal is transmitted. Figure B-10 shows an SSB system.

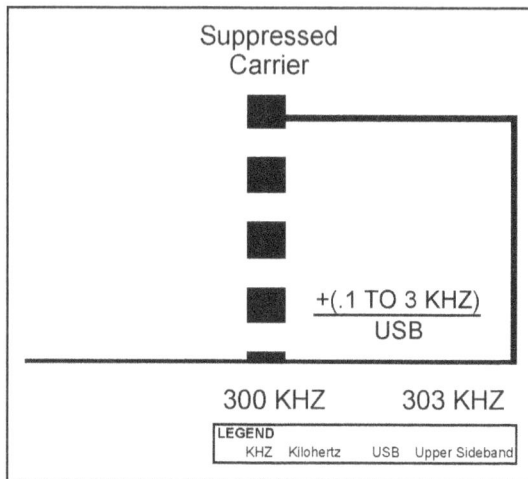

Figure B-10. SSB system

B-83. The upper side band is higher in frequency than the carrier and the lower side band is lower in frequency. Either sideband can be used for communications, provided both the transmitter and the receiver are adjusted to the same sideband. Most Army SSB equipment operates in the upper side band mode.

B-84. The transmission of only one sideband leaves open that portion of the RF spectrum normally occupied by the other sideband of an AM signal. This allows more emitters to be used within a given frequency range.

B-85. SSB transmission is used in applications where it is desired to—

- Obtain greater reliability.
- Limit size and weight of equipment.
- Increase effective output without increasing antenna voltage.
- Operate a large number of radio sets without heterodyne interference (whistles and squeals) from RF carriers.
- Operate over long ranges without loss of intelligibility due to selective fading.

This page intentionally left blank.

Appendix C

Antenna Selection

Merely selecting an antenna that radiates at a high elevation angle is not enough to ensure optimum communications. This appendix addresses the importance of HF, VHF and UHF antenna selection.

HIGH FREQUENCY ANTENNA SELECTION

C-1. The HF portion of the radio spectrum is very important to communications. Radio waves in the 3–30 MHz frequency range are the only ones that are capable of being reflected or returned to Earth by the ionosphere with predictable regularity. To optimize the probability of a successful sky wave communications link, select the frequency and take-off angle that is most appropriate for the time of day transmission is to take place.

C-2. Various large conducting objects, in particular the Earth's surface, will modify an antenna's radiation pattern. Sometimes nearby scattering objects may modify the antenna's pattern favorable by concentrating more power toward the receiving antenna. Often, the pattern alteration results in less signals being transmitted toward the receiver.

C-3. When selecting an antenna site, the operator should avoid as many scattering objects as possible. Although NVIS is the chief mode of short-haul HF propagation, the ground wave and direction (LOS) modes are also useful over short paths. How far a ground wave is useful depends on the electrical conductivity of the terrain or body of water over which it travels. The direct wave is useful only to the radio horizon, which extends slightly beyond the visual horizon.

ANTENNA SELECTION PROCEDURES

C-4. Selecting the right antenna for an HF radio circuit is very important. When selecting an HF antenna, first consider the type of propagation. Ground wave propagation requires low take-off angle and vertically polarized antennas. The whip antenna included with most radio sets provides good omnidirectional ground wave radiation.

C-5. Selecting an antenna for sky wave propagation is very complex. First, find the circuit (range) distance so that the required take-off angle can be determined. A circuit distance of 966 km (600 miles) requires a take-off angle of approximately 25 degrees during the day and 40 degrees at night. Select a high gain antenna (25–40 degrees). If propagation predictions are available, skip this step, since the predictions will probably give the take-off angles required.

C-6. Next, determine the required coverage. A radio circuit with mobile (vehicle) stations or several stations at different directions from the transmitter requires an omnidirectional antenna. A point-to-point circuit uses either a bidirectional or directional antenna. Normally, the receiving station location dictates this choice. Refer to Table C-1 for take-off angles versus distance.

Table C-1. Take-off angle versus distance

Take off Angle (Degrees)	Distance			
	F2 Region Daytime		F2 Region Nighttime	
	km	miles	km	miles
0	3220	2000	4508	2800
5	2415	1500	3703	2300
10	1932	1200	2898	1800
15	1450	900	2254	1400
20	1127	700	1771	1100
25	966	600	1610	1000
30	725	450	1328	825
35	644	400	1127	700
40	564	350	966	600
45	443	275	805	500
50	403	250	685	425
60	258	160	443	275
70	153	95	290	180
80	80	50	145	90
90	0	0	0	0

C-7. Before selecting a specific antenna, examine the available construction materials. At least two supports are needed to erect a horizontal dipole, with a third support in the middle for frequencies of 5 MHz or less. When support items are unavailable, the dipole cannot be constructed, and another antenna should be selected. Examine the proposed antenna site to determine if the antenna will fit the mission requirements. If not, select a different antenna.

C-8. The site is another important consideration. Usually, the tactical situation determines the position of the communications antenna. The ideal setting would be a clear, flat area (no trees, fences, power lines, or mountains). Unfortunately, an ideal location is seldom available. Choose the clearest, flattest area possible. Often, an antenna must be constructed on irregular sites. This does not mean that the antenna will not work. It means that the site will affect the antenna's pattern and function.

C-9. After selecting the antenna, determine how to feed the power from the radio to the antenna. Most tactical antennas are fed with coaxial cable (RG-213). Coaxial cable is a reasonable compromise of efficiency, convenience, and durability. Issued antennas include the necessary connectors for coaxial cable or for direct connection to the radio.

C-10. Problems may arise in connecting field expedient antenna. The horizontal half-wave dipole uses a balanced transmission line (open-wire). Coaxial cable can be used, but it may cause unwanted RF current.

C-11. A balun prevents unwanted RF current flow, which causes a radio to be hot or shock the RTO. Install the balun at the dipole feed point (center) to prevent unwanted RF current flow on the coaxial cable. If a balun is unavailable, use the coaxial cable that feeds the antenna as a choke. Connect the cable's center wire to one leg of the dipole and the cable braid to the other leg. Form the coaxial cable into a 6-inch coil (consisting of ten turns), and tape it to the antenna under the insulator for support.

DETERMINING ANTENNA GAIN

C-12. Figure C-1 shows the vertical antenna pattern for the 32 foot vertical whip antenna. The numbers along the outer ring (90, 80 and 70 degrees) represent the angle above the Earth; 90 degrees would be straight up, and 0 degrees would be along the ground. Along the bottom of the pattern are numbers from -10 (at the center) to = 15 (at the edges). These numbers represent the dBi over an isotropic radiator.

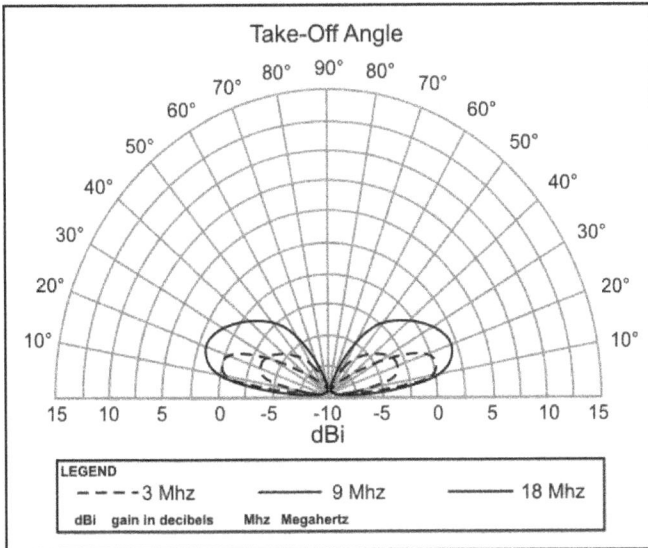

Figure C-1. 32-foot vertical whip, vertical antenna pattern

C-13. To find the antenna gain at a particular frequency and take-off angle, locate the desired take-off angle on the plot. Follow that line toward the center of the plot to the pattern of the desired frequency. Drop down and read the gain from the bottom scale. If the gain of 32 foot vertical whip at 9 MHz and 20 degree take-off angle is desired, locate 20 degrees along the outer scale. Follow this line to the 9 MHz pattern line. Move down to the bottom scale. The gain is a little less than 2.5 dBi. The gain of the 32 foot vertical whip at 9 MHz and 20 degrees is 2 dBi.

C-14. Once the antenna's overall characteristics are determined, use the HF antenna selection matrix (Table C-2) to find the specific antenna for a circuit. If the proposed circuit requires a short-range, omnidirectional, wideband antenna, the selection matrix shows that the only antenna that meets all the criteria is the AS-2259/GR.

Table C-2. HF antenna selection matrix

	Use				Directivity			Polarization		Bandwidth	
	Ground Wave	Skywave — Short 500 Miles	Skywave — Medium 500 to 1200 Miles	Skywave — Long 1200 Miles	Omnidirectional	Bidirectional	Directional	Horizontal	Vertical	Wide	Narrow
AS-2259/GR		X			X					X	
Vertical Whip	X				X				X	X	
Half-Wave Dipole		X	X			X		X			X
Long Wire	X		X	X		X	X	X	X		
Inverted L	X	X	X			X	X	X	X		X
Sloping V	X		X	X			X	X	X	X	
Vertical Half Rhombic	X		X	X			X		X	X	X

UHF AND VHF ANTENNA SELECTION

C-15. The VHF portion of the radio spectrum extends from 30–300 MHz and the UHF range reaches from 300–3,000 MHz (3 GHz). Both frequency ranges are extremely useful for short-range (less than 50 km or 31 miles) communications. This includes point-to-point, mobile, air-to-ground, and general purpose communications. Wavelengths at these frequencies ranges are considerably shorter than those in the HF range and simple antennas are much smaller.

C-16. Because VHF and UHF antennas are small, it is possible to use multiple radiating elements to form arrays, which provide a considerable gain in a given direction or directions. An array in an arrangement of antenna elements, usually dipoles, used to control the direction in which most of the antenna's power is radiated.

C-17. Within the VHF and UHF portion of the spectrum, there are sub-frequencies bands for specific uses such VHF aircraft band, UHF aircraft band and public communications. (Refer to FMI 6-02.70 for more information on spectrum management.)

POLARIZATION

C-18. In many countries, FM and television broadcasting in the VHF range use horizontal polarization. One reason is because it reduces ignition interference, which is mainly vertically polarized. Mobile communications often is vertical polarization or two reasons. First, the vehicle antenna installation has physical limitations, and second, so that reception or transmission will not be interrupted as the vehicle changes it's heading to achieve omnidirectionality.

C-19. Using directional antennas and horizontal polarization (when possible) will reduce manmade noise interference in urban locations. Horizontal polarization, however, should be chosen only where an antenna height of many wavelengths is possible. Ground reflections tend to cancel horizontally polarized waves at low angles. Use only vertically polarized antennas when the antenna must be located at a height of less than 10 meters (32.8 ft) above the ground, or where omnidirectional radiation or reception is desired.

GAIN AND DIRECTIVITY

C-20. VHF and UHF (above 30 MHz) antenna gain are extremely important for several reasons. Assuming the same antenna gain and propagation path, the received signal strength drops as frequency is increased. At VHF and UHF, more of the received signal is lost in the transmission line than is lost at HF. A 10–20 dB loss it not uncommon in a 30 meter (98.4 ft) length of coaxial line at 450 MHz.

C-21. At frequencies below 30 MHz, system sensitivity is almost always limited by receive noise rather than by noise external to the antenna. Generally, wider modulation or signal bandwidths are employed in VHF and UHF transmissions than at HF. Since system noise power is directly proportional to bandwidth, additional antenna gain is necessary to preserve a usable S/N ratio.

C-22. VHF and UHF antenna directivity (gain) aids security by restricting the amount of power radiated in unwanted directions. Receiver sensitivity is generally poorer at VHF and UHF (with the exception of high quality state-of-the-art receivers). Obstructions (buildings, trees, hills) may seriously decrease the signal strength available to the receiving antenna because VHF and UHF signals travel a straight LOS path.

C-23. Obtaining communications reliability over difficult VHF and UHF propagation paths requires considerable attention to the design of high-gain directive antenna arrays. Unlike HF communications, the shorter VHF and UHF wavelengths support walkie-talkie transceivers and simple mobile transmissions units. Communicating or receiving with such devices over distance beyond 1 or 2 km (.6 or 1.2 miles) requires maximum antenna gain at the base station or fixed end of the link.

C-24. Because VHF and UHF wavelengths are so short, reliability prediction of diffraction, refraction, and reflection effects are not practical. LOS paths must be entirely depended on. The best VHF and UHF communications are established with LOS paths that are free from obstacles. The VHF and UHF wavelengths are short enough that it is possible to construct resonant antenna arrays.

C-25. An array provides directivity (the ability to concentrate radiated energy into a beam that can be aimed at the intended receiver). Arrays of resonant elements, (half-wave dipoles, can be constructed of rigid metal rods or tubing or copper foil laid out or pasted on a flat non-conducting surface. Directing power helps to increase the range of the communications path and tends to decrease the likelihood of the interception of jamming from hostile radio stations. However, such highly directive antennas place an added burden on the RTO to ensure that the antenna is pointed properly.

ANTENNA PLANNING PROGRAMS

C-26. Several LOS radios require the planner/operator to do an analysis and prediction of the antennas LOS paths to ensure communications will be available from different planned locations. There are several programs designed to generate, store and disseminate communications information for antenna analysis and prediction. Several other programs can used to generate information even though it is not their primary purpose (such as ISYSCON [V]4/Tactical Internet Management System and Terrabase) The following paragraphs address several, but not all, programs that are available for use.

SYSTEM PLANNING, ENGINEERING AND EVALUATION DEVICE

C-27. The system planning, engineering, and evaluation device (SPEED) program is hosted by the Marine Corps Tactical Systems Support Activity. SPEED is a software package that provides communications planners with the tools necessary to engineer and plan radio communications analysis.

C-28. SPEED is a complete stand alone, self installing software package that provides the tools necessary to plan and analyze communications equipment. SPEED software contains HF analysis, radio coverage analysis, point-to-point, and satellite planning tools, which allows planning in response to rapidly changing communications architectures.

C-29. Communications planners will have to load topographical information before each operation to generate report, maps and matrices.

VOICE OF AMERICA COVERAGE ANALYSIS PROGRAM

C-30. Voice of America Coverage Analysis Program (VOACAP) software was released to the public and can be downloaded from the US Department of Commerce (National Telecommunications Information Administration/Institute for Telecommunications Sciences; Boulder, Colorado) to use as a HF prediction and analysis tool. VOACAP started as the Ionospheric Communications Analysis and Prediction (IONCAP) Program. (Voice of America is now organized as a component of the International Bureau of Broadcasting)

C-31. VOACAP offers the following capabilities—

- Easy to use graphical user interface.
- Detailed point-to-point graphs and area coverage maps for parameters such as:
    - S/N radio.
    - Required power gain.
    - Signal power.
    - MUF.
    - Take-off/arrival angle.
- Point-to-point performance versus distance for any given parameters at one or all user assigned frequencies.
- Calculates methods for antenna patterns.

C-32. Planners must input several parameters before VOACAP is capable of providing propagation prediction such as the method and the antennas used. Refer to http://www.its.bldrdoc.gov/elbert/hf.html for more information on VOACAP and Ionospheric Communications Enhanced Profile Analysis and Circuit (ICEPAC).

IONOSPHERIC COMMUNICATIONS ENHANCED PROFILE ANALYSIS AND CIRCUIT

C-33. ICEPAC is a full system performance model for HF radio communications in the frequency range of 2–30 MHz. capable of daily prediction methods with improved high latitude propagation models. ICEPAC is IONCAP with an ionospheric conductivity and electron density profile model added which is a statistical model of the large scale features of the northern hemisphere ionosphere. (For more information on ICEPAC refer to the article, "*Long–range Communications at High Frequencies.*")

Note. HFWIN 32 for Windows PC, ICEPAC and VOACAP are available at http://elbert.its.bldrdoc.gov/hf.html by the Department of Commerce.

Appendix D

Communications in Unusual Environments

Special consideration must be given to communications in unusual environments. This appendix addresses radio operations in cold weather, jungle, urban, desert, mountain areas, and nuclear areas.

COLD WEATHER OPERATIONS

D-1. SC radio equipment has certain capabilities and limitations that must be carefully considered when operating in extremely cold areas. However, in spite of significant limitations, the radio is still the normal means of communication in such areas.

D-2. One of the most important capabilities of radio in cold weather operations is its versatility. Vehicular mounted radios can easily be moved to almost any point where it is possible to install a command headquarters. Smaller, man packed radios can be carried to any point accessible by aircraft or on foot.

D-3. Interference by ionospheric disturbances limits radio communications in extremely cold areas. These disturbances, known as ionospheric storms, have a definite degrading effect on sky wave propagation. Moreover, both ionospheric storms and the Northern Lights (aurora borealis) activity can cause complete failure of radio communications; some frequencies may be blocked out completely by static for extended periods during storm activity. Fading, caused by changes in the density and height of the ionosphere, can also occur, and may last from several minutes to several weeks. These occurrences are difficult to predict, but when they do occur, the use of alternate frequencies, and a greater reliance on FM or other means of communications, is required.

TECHNIQUES FOR BETTER COMMUNICATIONS

D-4. Whenever possible, radio sets for tactical operations should be installed in vehicles, to reduce the problem of transportation and shelter for RTOs. This will help solve some of the grounding and antenna installation problems due to the climate.

D-5. It is difficult to establish good electrical grounds in extremely cold areas because of permafrost and deep snow. The conductivity of frozen ground is often too low to provide good ground wave propagation. To improve ground wave operation, use a counterpoise to offset the degrading effects of poor electrical ground conductivity. When installing a counterpoise, remember to install it high enough above the ground so it will not be covered by snow.

D-6. In general, antenna installation in arctic-like areas presents no serious difficulties. However, installing some antennas may take longer because of adverse working conditions. Tips for installing antennas in extremely cold areas include—

- The mast sections and antenna cables must be handled carefully since they become brittle in very low temperatures.
- Whenever possible, antenna cables should be constructed overhead to prevent damage from heavy snow and frost.
- Nylon rope guys, if available, should be used in preference to cotton or hemp, because nylon ropes do not readily absorb moisture, and are less likely to freeze and break.
- An antenna should have extra guy wires, supports, and anchor stakes to strengthen it, and to withstand heavy ice and wind loading.

D-7. Some radios (generally older generation radios) adjusted to a particular frequency in a relatively warm place, may drift off frequency when exposed to extreme cold; low battery voltage can also cause frequency drift. When possible, allow a radio to warm up several minutes before placing it into operation. Since extreme cold tends to lower output voltage of a dry battery, try warming the battery with body heat before operating the radio set; this minimizes frequency drift.

D-8. Flakes or pellets of highly electrically charged snow are sometimes experienced in northern regions. When these particles strike the antenna, the resulting electrical discharge causes a high-pitched static roar that can blanket all frequencies. To overcome this static, antenna elements can be covered with polystyrene tape and shellac.

MAINTENANCE IMPROVEMENT

D-9. The maintenance of radio equipment in extreme cold presents many difficulties. Radio sets must be protected from blowing snow because snow will freeze to dials and knobs, and will blow into the wiring to cause shorts and grounds. Cords and cables must be handled carefully since they may lose their flexibility in extreme cold. All radio equipment and power units must be properly winterized. Check the appropriate TM for winterization procedures. The following paragraphs provide suggestions for radio maintenance in arctic areas.

Power Units

D-10. As the temperature goes down, it becomes increasingly difficult to operate and maintain generators. Generators should be protected as much as possible from the weather.

Batteries

D-11. The effect of cold weather conditions on wet or dry cell batteries depends on the type of battery, the load on the battery, and the degree of exposure to cold temperatures. Batteries perform best at moderate temperatures and generally have a shorter life at very cold temperatures.

Shock Damage

D-12. Damage may occur to vehicular radio sets by the jolting of the vehicle. Most synthetic rubber shock mounts become stiff and brittle in extreme cold, and fail to cushion equipment. Check the shock mounts frequently, and change them as required.

Winterization

D-13. Check the TMs for the radio set and power source to see if there are special precautions for operation in extremely cold climates. For example, normal lubricants may solidify and permit damage or malfunctions to the radio equipment. They must be replaced with the recommended arctic lubricants. A light coat of silicon compound on antenna mast connections helps to keep them from freezing together and becoming hard to dismantle.

Microphones

D-14. Use standard microphone covers to prevent moisture from breath freezing on the perforated cover plate of the microphone. If standard covers are not available, improvise a suitable cover from rubber or cellophane membranes, or from rayon or nylon cloths.

Breathing and Sweating

D-15. A radio set generates heat when it is operated. When turned off, the air inside cools and contracts, drawing cold air into the set from the outside. This is called breathing. When a radio breathes and the still-hot parts come in contact with subzero air, the glass, plastic, and ceramic parts of the set may cool too rapidly and break.

D-16. Sweating occurs when cold equipment is brought suddenly into contact with warm air, moisture will condense on the equipment parts. Before cold equipment is brought into a heated area, it should be wrapped in a blanket or parka to ensure it will warm gradually to reduce sweating. Equipment must be thoroughly dry before it is taken back out into the cold air, or the moisture will freeze.

Vehicular Mounted Radios

D-17. These radios present special problems during winter operations because of their continuous exposure to the elements. Proper starting procedures must be observed. The radio's power switch must be off prior to starting the vehicle, especially when vehicles are slave-started. If the radio is cold soaked from prolonged shutdown, frost may have collected inside the radio and could cause circuit arcing. Hence, time should be allowed for the vehicle's heater to warm the radio sufficiently so that any frost collected within the radio has a chance to thaw.

D-18. The defrosting process may take up to an hour. Once the radio has been turned on, it should warm up for approximately 15 minutes before transmitting or changing frequencies; this allows components to stabilize.

D-19. If a vehicle is operated at a low idle with radios, heater, and lights on, the batteries may run down. Before increasing engine revolutions per minute to charge the batteries, radios should be turned off to avoid an excessive power surge.

OPERATIONS IN JUNGLE AREAS

D-20. Limitations on radio communications in jungle areas stem from the climate and the density of jungle growth. The hot and humid climate increases the maintenance problems of keeping equipment operable. Thick jungle growth acts as a vertically polarized absorbing screen for RF energy that, in effect, reduces transmission range. Therefore, increased emphasis on maintenance and antenna site selection is inherently important when operating in jungle areas.

D-21. Radio communications in jungle areas must be carefully planned, because dense jungle growth, heavy rains, and hilly terrain all significantly reduces the range of radio transmission. Trees and underbrush absorb VHF and UHF radio energy. In addition to the ordinary free space loss between transmitting and receiving antennas, a radio wave passing through a forest undergoes an additional loss. This extra loss increases rapidly as the transmission frequency increase. Near the ground (antenna heights of less than 3 meters [9.8 ft]) vertical polarization is preferred. However, if it is possible to elevate the receiving and transmitting antenna as much as 10–20 meters (32.8–65.6 ft), horizontal polarization is preferable to vertical polarization. Considerable reduction in total path loss results if either or both the transmitting and receiving antenna can be placed above the tree level through which communications must be made.

D-22. SC radios can be deployed in many configurations, especially man packed, which make it a valuable communications asset. The capabilities and limitations of tactical radios must be carefully considered when used by friendly forces in a jungle environment. The mobility and various configurations in which the tactical radio can be deployed are its primary advantages in jungle areas.

TECHNIQUES FOR BETTER COMMUNICATIONS

D-23. The site selection of the antenna is the main problem in establishing radio communications in jungle areas. Techniques to improve communications in the jungle include—

- Placing antennas in clearings on the edge farthest from the distant station, and as high as possible.
- Keeping antenna cables and connectors off the ground to lessen the effects of moisture, fungus, and insects. This also applies to all power and telephone cables.
- Using complete antenna systems, such as broadband and dipoles. They are more effective than fractional wavelength whip antennas.

- Clearing vegetation from antenna sites. If an antenna touches any foliage, especially wet foliage, the signal will be grounded.
- Using horizontally polarized antennas in preference to vertically polarized antennas because vegetation, particularly when wet, will act like a vertically polarized screen and absorb much of any vertically polarized signal.

MAINTENANCE IMPROVEMENT

D-24. Because of moisture and fungus, the maintenance of radio sets in tropical climates is more difficult than in temperate climates. The high relative humidity causes condensation to form on the equipment, and encourages the growth of fungus. RTOs and maintenance personnel should check the appropriate TMs for any special maintenance requirements. Techniques for improving maintenance in jungle areas include—

- Keeping the equipment as dry as possible and in lighted areas to retard fungal growth.
- Keeping all air vents clear of obstructions so air can properly circulate for cooling and drying of the equipment.
- Using moisture and fungus proofing paint, tape, or silicone grease to protect equipment after repairs, or when painted surfaces have been damaged or scratched.

EXPEDIENT ANTENNAS

D-25. Dismounted patrols, and units of company size and below, can greatly improve their ability to communicate in the jungle by using expedient antennas. While moving, users are generally restricted to using the short or long whip antennas that come with their manpack radios. However, when not moving, constructing and using an expedient antenna will allow users to broadcast farther, and to receive more clearly. An antenna that is not tuned or cut to the operating frequency is not as effective as the whips that are supplied with the radio. Circuits inside the radio load the whips properly so they are tuned to give maximum output. Whips are not as effective as a tuned doublet or a broadband (such as the OE-254), when the doublet or broadband is tuned to the operating frequency.

OE-254 Expedient Type Antenna

D-26. When used properly, the expedient OE-254 type antenna will increase the ability to communicate. In its entirety, the OE-254 type antenna is bulky and heavy, and is not generally acceptable for dismounted patrols or small unit operations. A Soldier can manage by, carrying only the masthead and antenna sections, mounting these on wooden poles, or hanging them up when not on the move.

OPERATIONS IN URBAN AREAS

D-27. Radio communications in urbanized terrain pose special problems. When the Army is engaged in urban combat operations the communications situation is considerably different from the situation faced by civil government or cell phone users. Military factors include—

- Restriction of operation to the frequency range of common military radios (2–512 MHz).
- Limits on the output power of military radio equipment.
- Limited number of available repeater assets if any.
- Limited access to good repeater locations due to enemy action.
- Need to communicate to both outside street locations and inside structures.
- Lack of standard compact antenna systems useful for urban combat.
- Severe restrictions on the movements of system users.
- Lack of manpower required to cover multiple signal sites can easily exceed available resources.
- Problems with obstacles blocking transmission paths.
- Problems with poor electrical conductivity due to pavement surfaces.
- Problem with commercial power lines interference.
- Distorted radio wave propagation in built-up areas and the limited availability of open lines of communication makes it difficult to move and install fixed station and multichannel systems.

D-28. FM and VHF radios that serve as the principle medium for C2 will have their effectiveness reduced in built-up areas. The operating frequencies and power output of these sets demand LOS between antennas. LOS at street level is not always possible in built-up areas. AM HF sets are less affected by the LOS problem because operating frequencies are lower, yet power output is greater. In past experiences, HF radios were not organic to the small units that conducted clearing operations; retransmitting VHF signals overcomes this limitation if available to utilize.

TECHNIQUES FOR BETTER COMMUNICATIONS

D-29. When available, wireless network extension stations in aerial platforms could provide the most effective means; depending on the requirement, organic wireless network extension sets will have to be used. Radio antennas should be hidden or blended in with the surroundings, so they will not be landmarks for the adversary to hone in on; water towers, commercial antennas, and steeples can conceal military antennas.

D-30. Wire can be laid while friendly forces are in static positions, but careful planning is necessary. Existing telephone poles can be used to raise wire lines above the streets. Ditches, culverts, and tunnels can be used to keep the wire below the streets. If these precautions are not taken, tracked and wheeled vehicles will constantly tear lines apart, and disrupt communications.

D-31. Messengers provide security and flexibility; however, once operations begin, messenger routes must be carefully selected to avoid any pockets of adversary resistance. Routes and time schedules should be varied to avoid establishing a pattern.

D-32. Pyrotechnics, smoke, and marker panels are also excellent means for communicating, but they must be well coordinated and fully understood by air and ground forces. The noise of combat in built-up areas makes it difficult to use sound signals effectively.

D-33. The possible seizure or retention of established communications facilities must be included in planning. Every effort should be made to prevent damage or destruction of these facilities. The local telephone system is already in place and tailored to the city or town. Army forces use local telephone systems to provide immediate access to wire communications with overhead and buried cable. This procedure helps overcome the problems encountered with radios, and provides a cable system less susceptible to combat damage.

D-34. Local media, such as newspapers, radio stations, and television stations, provide communications with the local populace after the level of combat declines. Additionally, intact police or taxi communications facilities are also possible radio systems, tailored to the city, with wireless network extension facilities already in place.

D-35. Radio equipped vehicles should be parked inside of buildings for cover and concealment when possible; dismount radio equipment, and install it inside buildings (in basements, if available). Place generators against buildings or under sheds to increase noise absorption and provide concealment and always remember to ensure adequate ventilation is available.

D-36. Another important consideration for urban combat is raw power. Obviously, the more power being used than the more path loss can be overcome and the deeper the signals will penetrate into structures. Common tactical VHF man-pack radios like SINCGARS have a maximum output power of four watts. The AN/PRC-150 I HF radio has a maximum output power of 20 watts. That is 7 dB more signal power to overcome losses caused by the path, path obstructions, inefficient antennas and other signal consuming factors. The extra power will help the radio but power relationships can be tricky, for example—

- 4 watts = 36 decibels above a miliwatt (dbm).
- 20 watts = 43 dbm.
- 50 watts = 47 dbm.
- 150 watts = 52 dbm.
- 400 watts = 56 dbm.

D-37. The dB is a logarithmic unit used to describe a ratio. The ratio may be power, voltage or intensity or several other factors but in this case it is power (watts). If the RTO looks at the math, he will see that he can measure the difference of two power levels by taking a logarithm of log $_{10}$ of their power ratio. If the ratio of power is, for example, two, meaning one radio transmitter is double the power of the other then the difference is 3dB. Put another way, for every 3dB gained by making a more efficient antenna system or cutting transmission line loss, is the equivalent to doubling the transmitter power.

D-38. The important point is that often, adjustments to antenna systems or operational frequencies to make an antenna more efficient can produce far more dBs of signal power than simply increasing the raw transmitter power. More power will always help overcome path loss for both NVIS and ground wave systems but many times it is not the best or only answer to the solution. If the radio is already operating at the maximum power that the transmitter can produce then these adjustments (to the antenna systems or frequencies) do become the only way to compensate for path loss and improve signal penetration in the urban combat environment.

> *Note.* It is important to remember that in some situations the power required to operate a radio may not need to be at the maximum power, use only the power necessary to operate.

D-39. Communications between two radio stations requires that the transmitter power-transmitter antenna gain-receiver antenna gain-receiver performance overcome the path loss between stations. A low power outstation radio such as a man-pack radio with an inefficient antenna used by forward troops can be "compensated for" to a degree when communicating with a base station that is typically using a higher performance receiver and a more efficient antenna. When the path is reversed, typically higher-power base-station transmitter and the more efficient antenna again compensates for lower performing combat unit radios in the net. Communications between low-power outstations is much more difficult and may even require wireless network extension through a more efficient base station.

D-40. In urban operations, small HF radios, such as the AN/PRC-150 I are extremely portable, but are antenna and power challenged based on location. A high degree of portable NVIS (sky wave) effect can be obtained when needed by simply physically reorienting standard vertical man-pack or vehicle (whip) antennas to the horizontal plane. Direct (surface wave) signals are simpler to generate and use inside structures are also produced from the same antenna by just leaving the antenna vertical.

High Frequency and Structures

D-41. Because of their longer wavelengths (lower frequency) HF (2–30 MHz) signals will naturally penetrate urban structures deeper than signals on higher, shorter wavelength frequencies. How deep the penetration depends on exact frequency, signal power level, antenna efficiency and the makeup of the urban structures in the path.

D-42. In all radio communications and particularly urban combat radio communications it is important to overcome path loss. The greater the radiated signal and the lower the frequency the more path loss can be overcome. This raises the probability of successful communications in urban areas and inside buildings.

D-43. As an example of HF signal penetration, it is not uncommon for a small ground penetrating radar transmitter operating in the HF frequency range to penetrate over 100 ft (30.4 meters) into common kinds of earth while the same power radar on a higher frequency will penetrate much less. So, if the RTO is using a common VHF military radio operating at 30 MHz (lowest frequency for SC ground-to-air radio systems) and replaces it with an HF radio AN/PRC-150 I operating at 5 MHz the path loss drops by 20 dB because of the way that longer wavelength (lower frequency) signals propagate. In this case lowering the frequency is the equivalent to increasing the power of the transmitter by a factor of almost seven.

OPERATIONS IN DESERT AREAS

D-44. Radios are usually the primary means of communications in the desert. They can be employed effectively in desert climate and terrain to provide the highly mobile means of communications demanded by widely dispersed forces. However, desert terrain provides poor electrical ground and counterpoises are needed to improve operation. The following paragraphs address operations in desert or arid areas.

D-45. Dust and extreme heat are two of the biggest problems involved in desert operations. Temperatures may vary from 58° Celsius (136° Fahrenheit), in summer, to -46° Celsius (-50° Fahrenheit), in winter. The heat can take a toll on generators, wire, communications equipment, and personnel.

D-46. Dust and sand particles damage equipment. Some CNRs have ventilating ports and channels that may clog with dust. These must be checked regularly, and kept clean to prevent overheating.

D-47. Grounding equipment in a desert environment is difficult, and can be accomplished by burying ground plates in the sand and frequently pouring salt solutions on them. Ensure equipment (for example, generators and air filters) is cleaned daily to prevent equipment damage.

TECHNIQUES FOR BETTER COMMUNICATIONS

D-48. It is essential that antennas be cut or adjusted to the length of the operating frequency. Directional antennas must be positioned exactly in the required direction; approximate azimuth produced by guesswork is not sufficient. A basic whip antenna relies on the capacitor effect, between itself and the ground, for efficient propagation. The electrical ground may be very poor, and the antenna performance alone may be degraded by as much as one-third if the surface soil lacks moisture (which is normally the case in the desert).

D-49. If a ground-mounted antenna is not fitted with a counterpoise (refer to Chapter 9 for more information on a counterpoise), the ground around it should be dampened using any fluid available. Vehicle mounted antennas are more efficient if the mass (main structure) of the vehicle is forward of the antennas, and is oriented toward the distant station.

D-50. Keep all radios cool and clean in accordance with preventive maintenance. Operate them in a shaded or ventilated area, and at low power whenever possible. Place a flat sheet of wood, cardboard or a vehicle's canvas top over the top of the radio to create manmade shade. Leaving a space between the wood/cardboard and the radio will help to further cool the radio by causing air to circulate in the shaded area between the radio and the wood. Using caution, cover hot radios with a damp cloth (ensure it is not soaking wet) without blocking air ventilation outlets; moisture evaporation from the cloth will also cool the radio.

D-51. Desert terrain can cause excessive signal attenuation, making planning ranges shorter. Desert operations require dispersion, yet the environment is likely to degrade the transmission range of radios, particularly VHFs (FM) fitted with secure equipment. This degradation is most likely to occur during the hottest part of the day, from approximately 1200–1700 hours.

D-52. If, during the hottest time of day, CNR stations begin to lose contact, alternative communications plans must be ready, and may include—
- Using relay stations, including an airborne relay station (the aircraft must remain at least 4,000 meters behind the line of contact). Ground relay stations or wireless network extension are also useful, and should be planned in conjunction with the scheme of maneuver.
- Deploying any unemployed vehicle with a radio as a relay between stations.
- Using alternative radio links, such as VHF multichannel telephones at higher level or HF-SSB voice.

D-53. After dark, rapid temperature drops will cause a heat inversion that can disrupt radio communications until the atmosphere stabilizes.

D-54. Generally, wire will not be used because military operations will be fluid; however, wire may be of some value in some static defensive situations. When possible, bury wire and cables deep in the soft sand to prevent heat damage to cable insulation, as well as vehicle, or foot traffic damage.

D-55. Prevent the exposure of floppy disks and computers to dust and sand. Covering computers and disks with plastic bags will reduce damage. However, extended periods of covering computers and/or radios may cause condensation inside these components and subsequent equipment damage or data loss. Compressed air cans will facilitate the cleaning of keyboards and other components of computer systems.

D-56. Wind-blown sand and grit will damage electrical wire insulation over a period of time. All cables that are likely to be damaged should be protected with tape before insulation becomes worn. Sand will also find its way into parts of items such as "spaghetti cord" plugs, either preventing electrical contact or making it impossible to join the plugs together. A brush, such as an old toothbrush, should be carried and used to clean such items before they are joined.

D-57. Static electricity is prevalent in the desert. It is caused by many factors, one of which is wind-blown dust particles. Extremely low humidity contributes highly to static discharges between charged particles. Poor grounding conditions aggravate the problem. Be sure to tape all sharp edges (tips) of antennas to cut down on wind-caused static discharges and the accompanying noise. If you are operating from a fixed position, ensure that equipment is properly grounded at all times. Since static-caused noise diminishes with an increase in frequency, use the highest frequencies that are available and authorized.

OPERATIONS IN MOUNTAIN AREAS

D-58. Radio operations in mountainous areas have some of the same problems as in cold weather areas. Mobility is difficult in mountainous terrain, and it can be difficult to find a level area for a communications site.

D-59. Generators and communications equipment need level ground to operate properly. It may difficult to drive ground rods and guy wire stakes into rocky, mountainous terrain and an alternate grounding method may be necessary. This rocky soil provides poor grounds; however, adding salt solutions will improve electrical flow.

TECHNIQUES FOR BETTER COMMUNICATIONS

D-60. When operating in mountainous terrain, additional wireless network extension assets will be needed. LOS paths are more difficult to plan, but use of relays improves communications. Positioning antennas is crucial in mountainous terrain, as moving an antenna, even a small distance, can drastically affect reception.

OPERATIONS IN A NUCLEAR AREA

D-61. A nuclear area will adversely affect sensitive radio equipment and components. Take measures to protect signal equipment, and ensure equipment survivability and availability for future use. Nearly everyone is aware of the effects of nuclear blast, heat, and radiation. The ionization of the atmosphere by a nuclear explosion will have degrading effects on communications because of static and the disruption of the ionosphere.

D-62. EMP is the radiation generated as a result of a nuclear detonation. Gamma rays, high energy photons, radiate outward from the point of the nuclear detonation, and strip electrons from the atoms in the air. This creates a wall of fast moving, negatively charged electrons which undergo rapid deceleration, radiating an intense electromagnetic field. This electromagnetic energy will affect unprotected communications equipment, causing disruption and/or destruction of delicate circuitry and components. The residual ionized cloud will also cause disruption of transmissions.

D-63. EMP has a great "killing range." EMP can disable electronic systems as far as 6,000 km (3,720 miles) (for an above the atmosphere [exoatmospheric] or high altitude EMP) from the site of the detonation. EMP can also cause severe disruption and sometimes damage when other weapon effects are absent. A high yield nuclear weapon, burst above the atmosphere, could be used to knock out a SC TACSAT communications system's operational status without doing any other significant damage. The range of EMP is diminished if the weapon is detonated at a lower altitude within the atmosphere.

D-64. An idea of the strength of EMP can be gained when we compare it with fields from man-made sources. A typical high level EMP could have an intensity (when taking into account the rise time, duration and amplitude of the pulse) which is one thousand times more intense than a radar beam. A radar beam has sufficient power to cause biological damage such as blindness or sterilization. The EMP spectrum is broad and extends from low frequencies into the UHF band. The most likely EMP effect would be stopping communications service temporarily. This can occur even without permanent damage. This delay could give an enemy enough of an advantage to change the outcome of the battle.

D-65. All TACSAT communications systems incorporate built-in features and techniques to counter the EMP effects. Shielding can further reduce the level of the EMP. Shielding is using equipment location and possible known directions of nuclear blasts to reduce EMP exposure. Shielding also depends on good grounding. Electronic systems depend on protection against EMP and signal equipment is very susceptible to EMP.

TECHNIQUES FOR BETTER COMMUNICATIONS

D-66. All equipment not required in primary systems should remain disconnected and stored within a sealed shelter, or other shielded enclosure, for protection from EMP. This reduces the likelihood of all equipment being simultaneously damaged by EMP, and provides a source of backup components to reinstall affected systems.

D-67. Wire and cable must be shielded and properly grounded. Keep the cable length as short as possible. Connect shields on all cables to the grounding systems, where provided. Effective grounding is a must to reduce the effects of EMP.

D-68. Antennas should be disconnected from radio sets when not in use, and operational nets should be reduced to a minimum. Most tactical radios with fully closed metal cases will provide adequate EMP protection if all external connectors have been removed. Placing radios in vehicles, vans, and underground shelters provides effective protection.

GENERAL RADIO SITE CONSIDERATIONS

D-69. The reliability of radio communications depends largely on the selection of a good radio site. Since it is difficult to select a radio site that satisfies all the technical, tactical, and security requirements, select the best site of all those available. In all cases, sites should be selected with the principals of site defense in mind—observation, avenues of approach, cover, obstacles, and key terrain.

D-70. Site selection is a leader and operator responsibility. It is also good planning to select both a primary site and an alternate site. If, for some reason, radio communications cannot be established and maintained at the primary location, the radio equipment can be moved a short distance to the alternate site.

D-71. A radio station must be located in a position that will assure communications with all other stations with which it is to operate, while maintaining a degree of physical and communication securities. To obtain efficiency of transmission and reception, the following factors should be considered—

- For operation at frequencies above 30 MHz, and whenever possible, select a location that will allow LOS communications. Try to avoid locations that provide the adversary with a jamming capability, visual sighting, or easy interception.
- Dry ground has high resistance, and limits the range of the radio set. If possible, locate the station near moist ground, which has much less resistance. Water, especially fresh water, greatly increases the distances that can be covered.
- Trees with heavy foliage absorb radio waves, and leafy trees have more of an adverse effect than evergreens. Keep the antenna clear of all foliage and dense brush. However, try to use available trees and shrubs for cover and concealment, and for screening from adversary jamming.

D-72. When located near man-made obstructions—

- Do not select an antenna position in a tunnel, or beneath an underpass or steel bridge. Transmission and reception under these conditions are almost impossible because of high absorption of RF energy.
- Avoid buildings located between radio stations, particularly steel and reinforced concrete structures; as they hinder transmission and reception. However, try to use buildings to camouflage antennas from the adversary.
- Avoid all types of suspended wire lines, such as telephone, telegraph, and high-tension power lines, when selecting a site for a radio station. Wire lines absorb power from radiating antennas located in their vicinity. They also introduce humming and noise interference in receiving antennas.
- Avoid positions adjacent to heavily traveled roads and highways. In addition to the noise and confusion caused by tanks and trucks, ignition systems in these vehicles may cause electrical interference.
- Do not locate battery charging units and generators close to the radio station.
- Do not locate radio stations close to each other.
- Locate radio stations in relatively quiet areas. The copying of weak signals requires great concentration by the RTO, and his attention should not be diverted by outside noises.

LOCAL COMMAND REQUIREMENTS

D-73. Radio stations should be located some distance from the unit headquarters or CP they serve. This distance separation will ensure that adversary DF capability will not target the CP with long range artillery fire, missiles, or aerial bombardment.

D-74. The locations selected should provide the best cover and concealment possible, consistent with good transmission and reception. Perfect cover and concealment may impair communications. The permissible amount of impairment depends upon the range required, the power of the transmitter, the sensitivity of the receiver, the efficiency of the antenna system, and the nature of the terrain. When a radio is being used to communicate over a distance that is well under the maximum range, some sacrifice of communications efficiency can be made to permit better concealment of the radio from adversary observation.

PRACTICAL CONSIDERATIONS

D-75. Manpack radio sets have sufficiently long cordage to permit operation from a concealed position (set and operator), while the antenna is mounted in the best position for communications. Some sets can be controlled remotely from distances of 30.4 meters (100 ft) or more. The remotely controlled set can be set up in a relatively exposed position, if necessary, while the RTO remains concealed.

D-76. All radio set antennas must be mounted higher than ground level to permit normal communications. Small tactical sets usually have whip antennas. These antennas are difficult to see from a distance, especially if they are not silhouetted against the sky. However, they have a 360 degree radiation pattern and are extremely vulnerable to adversary listening.

D-77. Avoid open crests of hills and mountains. A position protected from adversary fire just behind the crest gives better concealment and sometimes provides better communications. All permanent and semi-permanent positions should be properly camouflaged for protection from both aerial and ground observation. However, the antenna should not touch trees, brush, or the camouflage material.

D-78. Use one well-sited, broadband antenna and a FHMUX to serve several radios. This allows quicker set-up and disassemble times, and reduces camouflaging time and materials.

RADIO-TELEPHONE OPERATORS' SKILLS

D-79. The skills and technical abilities of the RTOs at the transmitter and receiver play important roles in obtaining the maximum range possible. The transmitter, output coupling, and antenna feeder circuits must be tuned correctly to obtain maximum power output. Additionally, both the radiating antenna and the receiving antenna have to be constructed properly with regard to both electrical characteristics and conditions of the local terrain. The RTO is the main defense against adversary interference. The skills of the RTO can be the final determining factor in maintaining C2 communications in the face of an adversary's efforts to disrupt it.

This page intentionally left blank.

Appendix E

Julian Date, Sync Time, and Time Conversion Chart

Accurate time is essential for SINCGARS to operate in the FH mode; a time variance greater than plus or minus four seconds will disrupt SINCGARS FH communications. This appendix addresses the Julian date, sync time, and Zulu time. It also provides a time zone conversion chart.

JULIAN DATE

E-1. The SINCGARS uses a special two-digit form of the Julian date as part of the sync time. The two-digit Julian date begins with 01 on 1 January and continues through 00, repeating as necessary to cover the entire year.

E-2. Since the two-digit Julian year terminates on 65 (or 66 for the leap year), every 1 January the Julian date must change to 01. This can be accomplished by—

- The NCS sending an ERF.
- Operators reloading time directly from an ANCD or PLGR.
- Operators manually changing the date in the radio by using the RT keypad.

E-3. Tables E-1 and E-2 show the two-digit Julian date calendars for regular and leap years, respectively.

Table E-1. Julian date calendar (regular year)

Day	JAN	FEB	MAR	APR	MAY	JUN	JUL	AUG	SEP	OCT	NOV	DEC
1	01	32	60	91	21	52	82	13	44	74	05	35
2	02	33	61	92	22	53	83	14	45	75	06	36
3	03	34	62	93	23	54	84	15	46	76	07	36
4	04	35	63	94	24	55	85	16	47	77	08	38
5	05	36	64	95	25	56	86	17	48	78	09	39
6	06	37	65	96	26	57	87	18	49	79	10	40
7	07	38	66	97	27	58	88	19	50	80	11	41
8	08	39	67	98	28	59	89	20	51	81	12	42
9	09	40	68	99	29	60	90	21	52	82	13	43
10	10	41	69	00	30	61	91	22	53	83	14	44
11	11	42	70	01	31	62	92	23	54	84	15	45
12	12	43	71	02	32	63	93	24	55	85	16	46
13	13	44	72	03	33	64	94	25	56	86	17	47
14	14	45	73	04	34	65	95	26	57	87	18	48
15	15	46	74	05	35	66	96	27	58	88	19	49
16	16	47	75	06	36	67	97	28	59	89	20	50
17	17	48	76	07	37	68	98	29	60	90	21	51
18	18	49	77	08	38	69	99	30	61	91	22	52
19	19	50	78	09	39	70	00	31	62	92	23	53
20	20	51	79	10	40	71	01	32	63	93	24	54
21	21	52	80	11	41	72	02	33	64	94	25	55
22	22	53	81	12	42	73	03	34	65	95	26	56
23	23	54	82	13	43	74	04	35	66	96	27	57
24	24	55	83	14	44	75	05	36	67	97	28	58

Table E-1. Julian date calendar (regular year) (continued)											
Day	JAN	FEB	MAR	APR	MAY	JUN	JUL	AUG	SEP	OCT	NOV	DEC
25	25	56	84	15	45	76	06	37	68	98	29	59
26	26	57	85	16	46	77	07	38	69	99	30	60
27	27	58	86	17	47	78	08	39	70	00	31	61
28	28	59	87	18	48	79	09	40	71	01	32	62
29	29		88	19	49	80	10	41	72	02	33	63
30	30		89	20	50	81	11	42	73	03	34	64
31	31		90		51		12	43		04		65

Table E-2. Julian date calendar (leap year)

Julian Date Calendar (Leap Year)												
Day	JAN	FEB	MAR	APR	MAY	JUN	JUL	AUG	SEP	OCT	NOV	DEC
1	01	32	61	92	22	53	83	14	45	75	06	36
2	02	33	62	93	23	54	84	15	46	76	07	37
3	03	34	63	94	24	55	85	16	47	77	08	38
4	04	35	64	95	25	56	86	17	48	78	09	39
5	05	36	65	96	26	57	87	18	49	79	10	40
6	06	37	66	97	27	58	88	19	50	80	11	41
7	07	38	67	98	28	59	89	20	51	81	12	42
8	08	39	68	99	29	60	90	21	52	82	13	43
9	09	40	69	00	30	61	91	22	53	83	14	44
10	10	41	70	01	31	62	92	23	54	84	15	45
11	11	42	71	02	32	63	93	24	55	85	16	46
12	12	43	72	03	33	64	94	25	56	86	17	47
13	13	44	73	04	34	65	95	26	57	87	18	48
14	14	45	74	05	35	66	96	27	58	88	19	49
15	15	46	75	06	36	67	97	28	59	89	20	50
16	16	47	76	07	37	68	98	29	60	90	21	51
17	17	48	77	08	38	69	99	30	61	91	22	52
18	18	49	78	09	39	70	00	31	62	92	23	53
19	19	50	79	10	40	71	01	32	63	93	24	54
20	20	51	80	11	41	72	02	33	64	94	25	55
21	21	52	81	12	42	73	03	34	65	95	26	56
22	22	53	82	13	43	74	04	35	66	96	27	57
23	23	54	83	14	44	75	05	36	67	97	28	58
24	24	55	84	15	45	76	06	37	68	98	29	59
25	25	56	85	16	46	77	07	38	69	99	30	60
26	26	57	86	17	47	78	08	39	70	00	31	61
27	27	58	87	18	48	79	09	40	71	01	32	62
28	28	59	88	19	49	80	10	41	72	02	33	63
29	29	60	89	20	50	81	11	42	73	03	34	64
30	30		90	21	51	82	12	43	74	04	35	65
31	31		91		52		13	44		05		66

SYNC TIME

E-4.   To maintain proper sync time, the SINCGARS uses seven internal clocks: a base clock, plus one for each of the six FH channels. Manual and cue settings will display the base clock time.

E-5.   With the fielding of the PLGR (and more recently the DAGR), all units were provided a ready source of highly accurate GPS time. By opening all nets on GPS time, and updating NCS RT sync time to GPS time daily, all nets of a division, corps, or larger force are continuously kept within the +/- four second window required for FH communications. Refer to TM 11-5820-890-7 for information on how to load the PLGR date and time into a SINCGARS.

ZULU TIME

E-6.   Zulu time remains in sync with the Naval Observatory Atomic Clock. Zulu time can be confirmed from the US Naval Observatory master clock telephone voice announcer Defense Switched Network (DSN) 762-1401, 762-1069 (Washington, DC) or DSN 560-6742 (Colorado Springs, Colorado). You can only connect to these numbers for a brief time before the call is terminated. If DSN is not available call (202) 762-1069 or (202) 762-1401. These are not toll-free numbers and callers outside the local calling area are charged at regular long-distance rates. Another alternative is to go to http://tycho.usno.navy.mil/, or use the time from a PLGR that is tracking at least one satellite. The NCS should update and verify net time daily or according to unit SOP.

TIME ZONE CONVERSIONS

E-7.   There are 25 integer World Time Zones from 12 through 0 Coordinated Universal Time (formerly Greenwich Mean Time) to +12. Each is 15 degrees longitude, as measured East and West, from the Prime Meridian of the earth at Greenwich, England.

E-8.   Table E-3 outlines each time zone around the world, and its relationship to Zulu time and Figure E-1 shows a world time zone map.

E-9.   When Coordinated Universal Time is 12:00, the diametrically opposed time zone is 00:00. This is indicated by the dashed line, and also indicates a date change. By convention, the area to the left of the dashed line is the following day, while the area to the right is the preceding day.

Table E-3. Example of world time zone conversion (standard time)

Y	X	W	V	U	T	S	R	Q	P	O	N	Z	A	B	C	D	E	F	G	H	I	K	L	M
Civilian Time Zones																								
IDLW	NT	HST	ASDT	PST	MST	CST	EST	AST	NST	AT	WAT	UTC	CET	EET	BT	ZP4	ZP5	ZP6	WAST	CCT	JST	GST	SBT	IDLE
1200	1300	1400	1500	1600	1700	1800	1900	2000**	2100	2200	2300	2400	0100	0200	0300	0400	0500	0600	0700	0800	0900	1000	1100	1200*

Standard Time=Universal Time + Value from Table

Z	0	E	+5	K	+10	P	-3	U	-8
A	+1	F	+6	L	+11	Q	-4	V	-9
B	+2	G	+7	M	+12	R	-5	W	-10
C	+3	H	+8	N	-1	S	-6	X	-11
D	+4	I	+9	O	-2	T	-7	Y	-12

* =Today   ** =Yesterday

AT-Azores Time
IDLW-International Date Line West
NST-Newfoundland Standard Time
HST-Hawaii Standard Time
EET-Eastern European Time
PST-Pacific Standard Time
MST-Mountain Standard Time
CST-Central Standard Time
EST-Eastern Standard time

AWST-Australian Western Standard Time
WAT-West Africa Time
UTC-Coordinated Universal Time
CET-Central European Time
IDLE-International Date Line East
BT-Baghdad
ZP-4
ZP-5
ZP-6

CCT-China Coast Time
GST-Guam Standard Time
JST-Japan Standard Time
ASDT-Alaska Standard Time
NT-Nome Time
WAST-West Africa Time Zone
AST-Atlantic Standard Time
SBT-Solomon Island Time

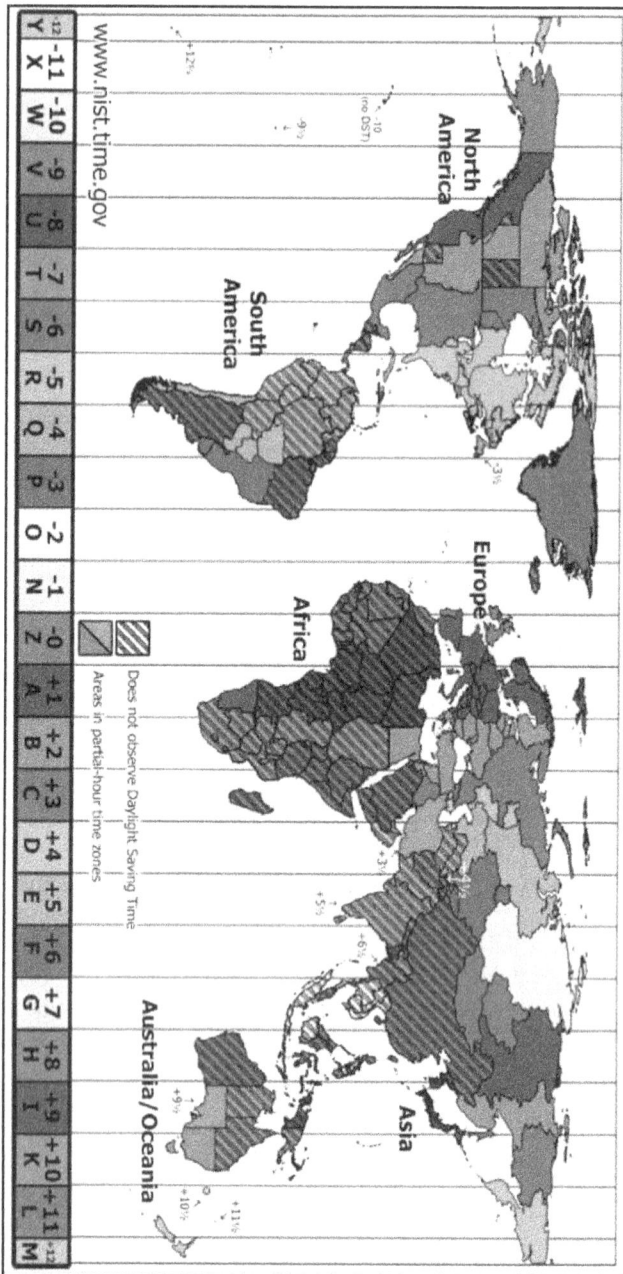

Figure E-1. World time zone map

Appendix F

Radio Compromise Recovery Procedures

Net compromise recovery procedures are essential to maintaining secure communications, and preventing an adversary from disrupting C2 communications due to loss or capture of COMSEC equipment. This appendix provides procedures for preventing and recovering a net after a compromise, and addresses recovery options available to the commander and his staff. This appendix is compliant with AR 380-40, and can be used as the core basis for a unit or taskforce SOP.

SECURE COMMUNICATIONS IMPERATIVES

F-1. The following imperatives will increase the unit's ability to operate without adversary intervention on its nets—

- ANCDs/SKLs below the battalion level (S-6) will only have the current TEK and KEK of the unit, and the minimum SOI data to perform the mission.
- ANCD loadsets will be loaded with NET ID 999 in each fill position, so not to compromise unit nets if captured. NET ID 999 will not be assigned as an operational net. (SINCGARS has the capability to manipulate all three digits of the NET ID.)
- ANCDs/SKLs and CIKs are always stored or transported separately to decrease ease of captured equipment operation by the adversary.
- Unique KEKs will be assigned down to the company level. (However, situations may arise that require unique KEKs at lower levels.)
- Units assign specific NET IDs as COMSEC recovery nets. (Predetermined NET IDs should be addressed in each unit's tactical SOP and/or OPORD.)

COMPROMISE DETERMINATION

F-2. The S-6, S-3, and S-2 will work together in determining the possibility of a compromise and the potential damage the compromise may cause. This damage is determined by evaluating what equipment was possibly captured or lost, and what COMSEC was loaded into the equipment. Upon determining there has been a compromise, COMSEC key replacement is required to secure the net.

F-3. Upon notification by the staff, the commander has three options. He can—

- Immediately implement the unit's compromise recovery procedures to secure the net.
- Extend the use of validated, intact COMSEC keys up to 24-hours. (Only if the commander is the controlling authority.) Commands must request permission to change COMSEC keys through the correct command channels.
- As a last resort, continue to use the compromised COMSEC keys.

COMPROMISE RECOVERY

F-4. If the controlling authority decides to continue using the compromised key, the commander, under advisement from the S-6/G-6 and staff, may initiate actions to protect net security.

F-5. If an operational radio and/or a filled ANCD/SKL falls into an adversary's hands, the unit SOP should assume the adversary has English-speaking Soldiers who can operate the radio and ANCD/SKL. The SOP should also assume the adversary is able to listen to US secure FH net communications and can transmit on that same US net, if desired.

F-6. Other assumptions and factors to consider if faced with a compromise recovery requirement include—

- Can the adversary move the captured radio and continue to operate that radio?
- What is the range of the captured radio?
- What is the expected duration of the battery or other power source?
- How long until the next periodic COMSEC update?
- How serious is the adversary's access to your net?
- What is the potential impact of the captured loadset on other nets?
- What was the nature of, and how critical is, the unit operation at the time that the compromise recovery was considered?

F-7. Two sets of net compromise recovery procedures exist to provide units guidance on recovering from a net compromise. Table F-1 provides procedures for those units that have compromised TEKs and KEKs, and Table F-2 provides procedures for those units that have compromised TEKs only. These procedures offer ways to help protect net security; however, this is not a substitute for distributing new COMSEC keys as soon as operationally possible.

Table F-1. Compromised net recovery procedures: compromised TEKs and KEKs

Step	Procedure
1	The NCS is advised of loss of radio and/or ANCD/SKL.
2	The S-6 notifies next higher command and/or controlling authority, and requests permission to change to reserve TEK.
3	The G-6/S-6 and the commander determine if compromise recovery action is warranted. Depending on the operational situation, the G-6/S-6 and the commander may elect to temporarily continue to use the presumably compromised net until it is determined that the compromise and compromise procedures will not interfere with current operations.
4	If compromise recovery action is required, the NCS broadcasts unit code word, meaning "Standby for activation of compromise procedures." (Adversary does not know the meaning of this code-word.)
5	In accordance with compromise procedures, each operator in the net will answer back with "WILCO, out," verifying they understand and will comply. The operator will then switch to the unit's predetermined alternate NET ID, and wait for the NCS to perform a net call.
6	The NCS maintains a tracking chart to log all subscribers confirming the code word. If possible, the NCS should maintain additional SINCGARS on the old NET ID to ensure that all users are moved to the alternate NET ID. (This is commonly called straggler control.)
7	The NCS then changes to the predetermined alternate NET ID and performs a net call. NCS operator logs in the users as they answer on the alternate NET ID.
8	Upon gaining controlling authority approval to change to the new TEK, the NCS will initiate a net call and inform all users of the manual COMSEC distribution plan. Each radio and ANCD/SKL will have to be manually filled from another device with the new COMSEC. (This is a mandatory physical distribution due to the KEK compromise.)
9	Upon complete distribution of the new COMSEC, the NCS will initiate a net call, informing the unit of the time to change to the new COMSEC, and return to the original NET ID.
10	At the designated time, the NCS will return to the original NET ID and log all subscribers on a tracking chart as they return to the original NET ID on the new COMSEC. If possible, the NCS should maintain an additional radio on the alternate NET ID to ensure that all users are moved over to the original NET ID.
11	The losing unit/net has now effectively recovered from the actual or potential compromise situation.

Table F-2. Compromised net recovery procedures: compromised TEKs

Step	Procedures
1	The NCS of the net is advised of loss of radio and/or ANCD/SKL.
2	The S-6 notifies next higher command and/or controlling authority, and requests permission to change to the reserve TEK.
3	The G-6/S-6 and the commander determine if compromise recovery action is warranted. Depending on the operational situation, the G-6/S-6 and the commander may elect to temporarily continue to use the presumably compromised net until they determine the compromise and compromise procedures will not interfere with current operations.
4	If compromise recovery action is required, the NCS broadcasts unit code-word, meaning "Standby for activation of compromise procedures." (Adversary does not know the meaning of this code-word, and does not know the alternate NET ID.)
5	In accordance with compromise procedures, each operator in the net will answer back with "WILCO, out," verifying that he understands and will comply. The operator will then switch to the alternate NET ID, and wait for the NCS to perform a net call.
6	The NCS maintains a tracking chart to log all subscribers confirming the code-word. If possible, the NCS should maintain an additional radio on the old NET ID to ensure that all users are moved over to the alternate NET ID. (This is commonly called straggler control.)
7	The NCS then changes to the predetermined alternate NET ID and performs a net call. NCS logs in users as they answer on the alternate NET ID.
8	Upon gaining approval from the controlling authority to change to the new TEK, the NCS will initiate a net call and OTAR procedures, or initiate a manual rekeying of the unit's SINCGARS and fill devices. (OTAR—automatic key procedures should only be used at the effective time of the COMSEC key.)
9	Upon complete distribution of the new COMSEC, the NCS will initiate a net call informing the unit of the time to change to the new COMSEC and return to the original NET ID.
10	At the designated time, the NCS will return to the original NET ID and log all subscribers on a tracking chart as they return to the original NET ID on the new COMSEC. If possible, the NCS should maintain an additional radio on the alternate NET ID to ensure that all users are moved to original NET ID.
11	The losing unit/net has now effectively recovered from the actual or potential compromise situation.

F-8.   Since the entire division/brigade is operating on the same TEK, the divisions/brigade G-6 may elect to have all nets change to a new TEK. If so, this change may be accomplished by the physical transfer from ANCD/SKL to ANCD/SKL, or by OTAR, as most appropriate for the operational situation.

This page intentionally left blank.

Appendix G

Data Communications

This appendix addresses data communications elements such as binary data, baud rate, modems and FEC.

BINARY DATA

G-1. Bits are part of a numbering system (binary digits) having a base of two that uses only the symbols 0 and 1. Thus, a bit is any variable that assumes two distinct states. For example, a switch is open or closed; a voltage is positive or negative. In terms of communications, words become binary digits for transformation over a channel (specific frequency range), via a HF radio transmitter, to a HF receiver.

G-2. A simple way to communicate binary data is to switch a circuit on and off in patterns that are interpreted at the other end; the same as the telegraph. Later schemes used a bit to select one of two possible states of the properties that characterize a carrier, FM or AM. Currently, the carrier assumes more than two states, and is able to represent multiple bits.

BAUD RATE

G-3. Data transmission speed is commonly measured in bps. Sometimes the word baud is used to represent bps, although the terms are different. Baud measures the signaling speed and is a measurement of symbols per second that are being sent. Symbols may represent a bit or more.

G-4. The bandwidth determines the maximum baud rate on a radio channel; the larger the bandwidth, the greater the baud rate. The rate at which information is transmitted (the bit rate) depends on how many bits are used per symbol.

ASYNCHRONOUS AND SYNCHRONOUS DATA

G-5. The transmission of data occurs in either the asynchronous or synchronous mode. In asynchronous data transmission, each character has a start and stop bit. The start bit prepares the data receiver to accept the character. The stop bit brings the data receiver back to the wait state. Synchronous data transmission eliminates the start and stop bits. This type of system typically uses a preamble (a known sequence of bits at the start of the message) to synchronize the receiver's internal clock and to alert the data receiver that a message is coming.

G-6. Asynchronous systems eliminate the need for complex synchronization circuits, but at the cost of higher overhead than synchronous systems. With asynchronous systems the start and stop bits increase the length of the character from 8 bits (one byte) to 10 bits, a 25 percent increase.

HIGH FREQUENCY MODEMS

G-7. The average voice radio cannot transmit data directly. Data digital voltage levels must be converted to audio using a modulator device that applies the audio to the transmitter. At the receiver, a demodulator converts the audio back to digital voltage levels. HF modems fall into three basic categories—

- Modems with slow-speed audio FSK capable of operating at data rates of 75, 150, 300 and 600 bps.
- High-speed parallel tone.
- High-speed serial tone capable of operating at data rates between 75 and 2, 400 bps.

G-8. The simplest modems use FSK to encode binary data. The input to the modulator is a digital signal that takes one of two possible voltage levels. The output of the modulator is an audio signal that is one of two possible tones. HF FSK systems are limited to data rates less than 75 bps, due to the effects of multipath propagation. Higher rates are possible with multitone FSK, which uses a greater number of frequencies.

G-9. High-speed HF modem technology, using both parallel and serial tone waveforms, allows data transmissions at up to 4,800 bps. The serial tone modem carries information on a single audio tone. This vastly improves data communications on HF channels, including greater toughness, less sensitivity to interference, and a higher data rate with more powerful FEC.

IMPROVED DATA MODEM

G-10. The improved data modem will allow both air and ground forces to exchange complex information in short bursts. It will permit simultaneous transmit/receive information from four different radios, interface with MIL-STD 1553 data bus, transmit data at 16,000 bps, and process messages up to 3,500 characters in length.

FORWARD ERROR CODING

G-11. FEC adds redundant data to the data stream to allow the data receiver to detect and correct errors. It does not require a return channel for the acknowledgment. If a data receiver detects an error, it simply corrects it and accurately reproduces the original data without notifying the data sender that there was an error.

G-12. FEC coding is most effective if errors occur randomly in a data stream. However, the HF medium typically introduces errors that occur in bursts. To take advantage of the FEC coding technique, interleaving randomizes the errors that occur in the channel. At the demodulator, de-interleaving reverses the process.

G-13. Soft-decision decoding further enhances the power of the error correction coding. In this process, a group of detected symbols that retains its analog character is compared against a set of possible transmitted code words. The system remembers the voltage from the detector, and applies a weighing factor to each symbol in the code word before making a decision about which code word was transmitted.

G-14. Data communications techniques are also used for encrypting voice calls by a VOCODER, a derivative of voice coder-decoder. The VOCODER converts sound into a data stream for transmission over a HF channel. The VOCODER at the receiving end reconstructs the data into telephone quality sound.

G-15. In addition to error correction techniques, high-speed serial modems may include two signal-processing schemes that improve data transmission. An automatic-channel equalizer compensates for variations in the channel characteristics as data is being received. An adaptive excision filter seeks output, and suppresses narrowband interference in the demodulator input, thereby reducing the effects of co-channel interference; interference on the same channel being used.

Appendix H

Co-Site Interference

Co-site interference is the effect of unwanted energy, due to emissions, radiation, or induction, on reception in a radio communications system. This could cause system performance degradation, misinterpretation, or loss of information. As telecommunication systems become more complex and several antennas are placed on the same platform, or when multiple radios in the same or dissimilar frequency bands are integrated within mobile communications CP platforms, interference becomes significant in system performance. This appendix addresses SINCGARS implications and co-site interference mitigation.

SINCGARS IMPLICATIONS

H-1. Due to SINCGARS FH capabilities, frequency management alone does not reduce co-site interference. The addition of computer central processing units, displays, switches, routers, hubs, and cables in the confined CP amplifies the potential for co-site interference.

H-2. Within a CP or a mobile platform (vehicle or aircraft), co-site interference depends on several factors, including—

- The number of transmitters within the restricted area.
- The duty cycle of each transmitter—the transmitting time of the radio, divided by the transmitting time plus the time before the next transmission. (Example: if a radio transmits for four seconds and waits six seconds before the next transmission, the duty cycle is 40 percent.)
- The hopset bandwidth (if hopping).
- An increase in the system data rate increases the electromagnetic flux of the system, thus increasing interference potential.
- Antenna placement.
- Equipment shielding.
- Bonding.
- Grounding.

H-3. SINCGARS that habitually transmit to distances of 35–40 km (21.7–24.8 miles), by themselves, can transmit at distances reduced to less than 5 km (3.1 miles) when influenced by co-site interference. This degradation, if not properly addressed, will adversely distress the flow of C2 communications. This distress may lead to the physical shutdown of non-critical systems that pass information onto critical systems.

H-4. Figure H-1 shows the mobile CP antenna configuration. (The antennas have been removed to avoid clutter). This mobile CP contains multiple radio systems, including FH SINCGARS, multiplexed on a single antenna within the CP. The close proximity and number of simultaneous transmitters produce unwanted emissions and degrade or block outstation receiver communications.

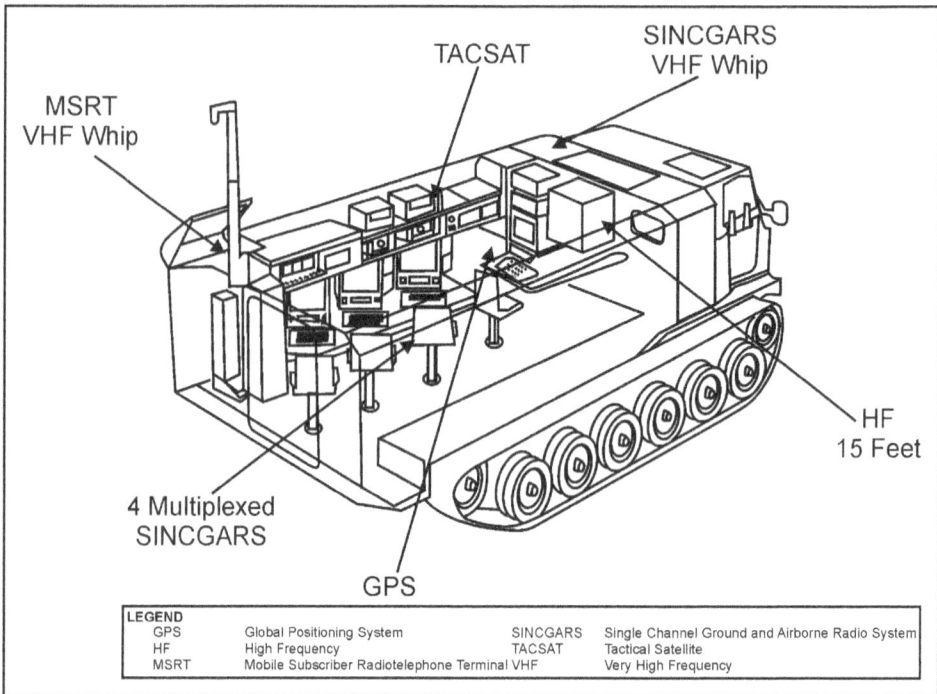

Figure H-1. Mobile command post antenna configuration

H-5. When a SINCGARS transmits at maximum power, a collocated mobile subscriber radiotelephone terminal (MSRT) radio cannot establish a link into the MSE area communications system. Antennas require 20+ ft of separation to overcome the SINCGARS-generated increase in background noise. This separation allows an acceptable S/N ratio for MSRT to establish a successful link.

H-6. If SINCGARS transmits at a power of four watts or less, the MSRT can effectively establish a voice link with some reduction in data quality. SINCGARS low power (4 watts) output reduces the SINCGARS planning range by 90 percent, and subjects the SINCGARS to increased noise generated by the collocated, transmitting MSRT system.

H-7. If SINCGARS is configured to hop outside the MSRT frequency range (59–88 MHz outside the Continental United States [OCONUS] or 40–50 MHz CONUS), plus an additional 5 MHz cushion in both areas of operation, the MSRT is relatively resistant to SINCGARS co-site interference. However, this causes a significant reduction of the available frequency spectrum, and a constraint on the capabilities of the SINCGARS. Full frequency range and full power hopping transmissions from SINCGARS will reduce MSRT operational distances by 94 percent. MSRT transmissions (16 watts) will degrade a co-sited SINCGARS operational planning distance by 74 percent. For all intents and purposes, full power operations of both systems can render them inoperable in many tactical situations.

CO-SITE INTERFERENCE MITIGATION

H-8. A number of options are available to mitigate co-site interference, but there are no comprehensive solutions. The user must decide if an option is applicable to his tactical situation, and take the appropriate action to resolve co-site interference.

H-9. Some equipment systems are not as critical as others. The S-6/G-6 must recommend to the commander a system priority list that ensures the transmission of critical mission information. During interference, the S-6/G-6 must be prepared to shut down less critical systems. The following paragraphs address ways to reduce co-site interference.

TRANSMISSION

H-10. When possible and operationally acceptable, transmit at the lowest power level. This allows collocated SINCGARS and MSRT antenna systems to operate with minimal interference in both data and voice communications at the receivers. This option may be unacceptable due to the significant transmission range reduction of the SINCGARS.

H-11. Remoting antennas and transmitting from the CP at low power to a full power wireless network extension system mitigates co-site interference. Certain critical TOC nets would then be able to maintain their high power advantage.

ANTENNA PLACEMENT

H-12. Antenna placement is critical when the antennas operate in the same or nearby frequency range(s); separate antennas as much as possible. The greater the separation between the transmitting and receiving antennas, the less interference encountered. TOCs could be issued a significant quantity of mast-mounted antennas (OE-254 or equivalent) to match the number of installed SINCGARS. Extra length low-loss coaxial transmission lines should be included with each requirement. However, this may cause an increase in the physical size of the CP location, and an increase in CP setup and disassembly times. Figure H-2 shows an example of proper antenna separation for an armored TOC.

H-13. Tilting the tops of the transmitting and receiving antennas away from each other can enhance vertically polarized ground wave communications. Tilt angles between 15 and 30 degrees will provide the best results; the best angle is achieved by trial and error.

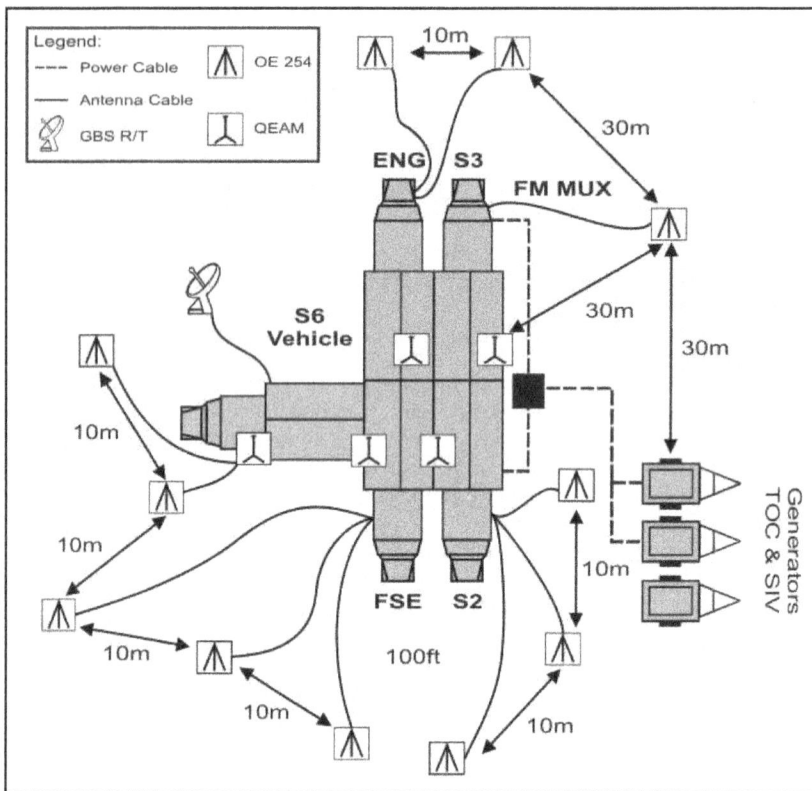

Figure H-2. Example of proper antenna separation for an armored TOC

DIRECTIONAL ANTENNAS

H-14. Use directional antennas whenever possible. This may require the prefabrication of VHF directional antennas, since these are not available in the current Army inventory. Change antenna polarization on systems where distance is not an issue. A horizontally polarized ground wave will have less signal loss than a vertically polarized ground wave if antenna heights exceed treetop levels or other horizontal energy absorbers.

MAST ASSEMBLIES

H-15. If possible, stack antennas in the null space of another vertical antenna. The radiation pattern of a vertical antenna has a deep energy void directly overhead (90 degrees). Figure H-3 shows possible antenna stacks. These mast assemblies would be configured to mount two OE-254 broadband antennas using vertical separation, as shown in configuration A of Figure H-3. Mast assemblies could also use a new design that incorporates the omnidirectional antennas into mast sections, as shown in configuration B of Figure H-3.

H-16. Both dual-antenna mast assemblies must provide at least 12 dB or greater antenna isolation (at 30 MHz) over that obtained using the same distance horizontal separation. Taking advantage of the lateral wave propagation of vertical antennas. Energy transference is negligible on a receiving antenna in this null space. Early fabrication of mounting devices may be required to achieve antenna stacking.

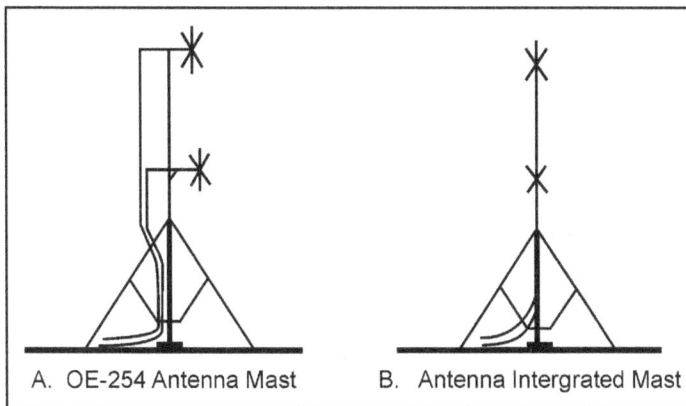

A. OE-254 Antenna Mast B. Antenna Intergrated Mast

Figure H-3. Possible antenna stacks

GROUNDING

H-17. Ensure electronic equipment within the CP is properly grounded. Proper grounding ensures that each item does not develop interference-producing electromagnetic fields, or simulate the properties of an unwanted, energy-radiating transmitting antenna within the CP.

H-18. Another option is to counterpoise the antenna. The wires used in the counterpoise should be either a half wavelength, or a full wavelength long for best results. A greater direction gain can be achieved by placing the counterpoise wires in the direction of the receiving antenna. (Refer to Chapter 9 for more information on how to construct a counterpoise.)

SINGLE-CHANNEL OPERATIONS

H-19. When operating against less sophisticated adversaries, using SINCGARS SC mode of operation also mitigates co-site interference. Even when operating at full power, properly chosen frequencies can reduce co-site interference, and provide increased range capability due to better bit error rate, inherent with SC operation.

INITIATIVES

H-20. Two co-site mitigation initiatives are the TD-1456/VRC, FHMUX, and the JTRS family of radios. Communications integration and co-site mitigation science and technology objectives products enhance both initiatives.

TD-1456/VRC, FHMUX

H-21. The FHMUX is a hardware solution to co-site interference. It is compatible with the SINCGARS in EP (FH) and SC (non-FH) modes of operation. Figure H-4 shows the FHMUX. The FHMUX is an antenna multicoupler that—

- Reduces visual signature of the command vehicle, by reducing the antenna count, thus increasing the survivability of the vehicle on the battlefield.
- Reduces collocated net-to-net interference (co-site).
- Reduces setup time for C2. The user erects one OE-254 antenna, and four nets are operational via the FHMUX. FHMUX is compatible with high power whip antennas, such as the AS-3900A/VRC or AS-3916/VRC.

- Reduces the parasitic effect of the antennas. The transmit radiation of one antenna in close proximity (10 ft/3 meters) will interact with another antenna producing undesirable distortions within the pattern of each antenna.
- Provides up to 300 meters (.3 km) multicoupler to antenna separation, to reduce exposure of the CP to hostile fire.
- Provides frequency conflict arbitration software that optimizes the transmission range. Table H-1 shows the effects of multiple transmitters (within a C2 vehicle) on transmission ranges with and without the FHMUX.

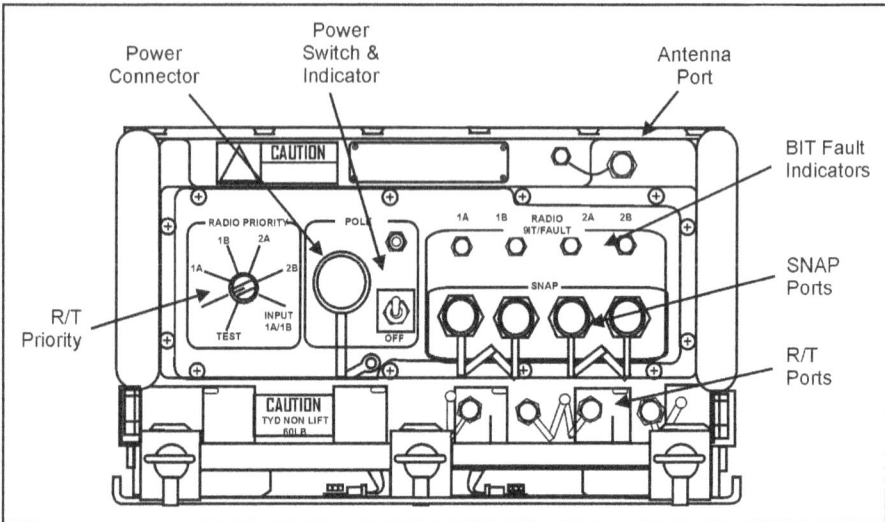

Figure H-4. Frequency hopping multiplexer

Table H-1. Transmitters and transmission ranges with and without the FHMUX

Transmitters On	Range to target receiver without FHMUX	Range to target receiver with FHMUX
zero	35 km/21.7 miles	35 km/21.7 miles
one	14 km/8.6 miles	32 km/19.8 miles
two	9 km/5.5 miles	27 km/16.7 miles
three	3 km/1.8 miles	19 km/11.8 miles

Note. Range to target receiver from a C2 vehicle, compared to the number of transmitters operating on the C2 vehicle.

H-22. The FHMUX contains bandpass filters that tune synchronously with the radios. These filters remove most of the broadband transmit interference. Signals coming from the antenna also pass through these bandpass filters, and strong, non-bandpass signals are removed. This greatly improves the performance of the radio system when in a co-site environment.

H-23. In the EP mode, the FHMUX is most effective when the hopset contains at least 800 channels and it is spread over at least 20 MHz of the VHF band. When enemy intrusion is not an issue, and the SC mode is used, the FHMUX is most effective when frequencies are separated by a five percent delta for each radio. (Refer to TM 11-5820-890-10-8 and TM 11-5820-890-23&P for more information on the FHMUX.)

Joint Tactical Radio System

H-24. The JTRS may eliminate most, if not all, co-site interference problems that occur when multiple radios in the same or dissimilar frequency bands are integrated within the same mobile communications CP platform. The JTRS operates at full performance levels, and does not degrade mission effectiveness of host systems/platforms engaged in their tactical environments, including weapons firing and movements.

H-25. New efforts, in conjunction with the JTRS, include the development of a VHF/UHF multiplexer, utilizing RF signal combining, and co-site mitigation technology to reduce the platform's antenna visual signature and JTRS self-jamming interference. The initial and objective multiplexer development efforts will exploit emerging technology applications in the areas of wideband interference mitigation and compact delay lines.

H-26. A new communications integration and co-site mitigation science and technology objective initiative includes a multiband VHF/UHF PA that will eliminate dissimilar legacy radio amplifiers and their logistics, training, and maintenance infrastructures, and will provide a modular, programmable JTRS waveform capability. The initial and objective PA development efforts will use laterally diffused metal oxide semiconductor and silicone carbide device technology to meet higher power and frequency requirements.

This page intentionally left blank.

Glossary

The glossary lists acronyms and terms with Army, multi-service, or joint definitions, and other selected terms. Where Army and joint definitions are different, *(Army)* follows the term. Terms for which FM 6-02.53 is the proponent manual (the authority) are marked with an asterisk (*). The proponent manual for other terms is listed in parentheses after the definition.

SECTION I – ACRONYMS AND ABBREVIATIONS

2G	second generation
3G	third generation
A&L	administrative and logistics
A2C2S	Army Airborne Command and Control System
ABCS	Army Battle Command System
ACES	automated communications engineering software
ACMES	Automated Communications Security Management and Engineering System
ADA	air defense artillery
AFATDS	Advanced Field Artillery Tactical Data System
AIRP	Army interference resolution program
AIRSIP	airborne system improvement program
AIS	Automated Information Systems
AKMS	Army Key Management System
ALE	automatic link establishment
AM	amplitude modulation
ANCD	automated net control device
ANDVT	advanced narrowband digital voice terminal
ANR	active noise reduction
AO	area of operations
AOR	area of responsibility
APCO	Association of Public Safety Communications Officials
AR	Army regulation
ARNG	Army National Guard
ARNGUS	Army National Guard of the United States
ASIP	advanced system improvement program
ASIP-E	advanced system improvement program-enhanced
ASCC	Army Service component commander
ATC	air traffic control
AWACS	Airborne Warning and Control System
BCT	brigade combat team

BFT	Blue Force Tracking
BIT	built-in test
BLOS	beyond line of sight
bps	bits per second
C2	command and control
CA	Civil Affairs
CDR	commander
CDU	control display unit
CEOI	communications-electronics operating instructions
CIK	cryptographic ignition key
CJCSI	Chairman Joint Chiefs of Staff instruction
CNR	combat net radio
CNRS	communications networking radio subsystem
CO	company
COA	course of action
COMSEC	communications security
CONUS	continental United States
COOP	Continuity of Operations Plan
COTS	commercial off-the-shelf
CP	command post
CPC	combat survivor evader locator planning computer
CREW	Counter Remote Control Improvised Explosive Device Warfare
CSEL	combat survivor evader locator
CSMA	carrier sense multiple access
CT	cipher text
CW	continuous wave
DA	Department of the Army
DAGR	defense advanced global positioning system receiver
DAMA	demand assigned multiple access
DAP	dynamically allocated permanent (virtual circuit)
dB	decibel
dBi	gain in decibels
dbm	decibels above a milliwatt
DF	direction finding
DOD	Department of Defense
DRA	data rate adapter
DTD	data transfer device
DSN	Defense Switched Network
DTE	data terminal equipment
EA	electronic attack
ECCM	electronic counter-countermeasures

EDM	enhanced data mode
EEFI	essential elements of friendly information
EHF	extremely high frequency
EKMS	Electronic Key Management System
e-mail	electronic mail
EMI	electromagnetic interference
EMP	electromagnetic pulse
EMSO	electromagnetic spectrum operations
ENM	Enhanced Position Location Reporting System network manager
EP	electronic protection
EPLRS	Enhanced Position Location Reporting System
ERF	electronic remote fill
ES	electronic warfare support
ESB	expeditionary signal battalion
ESIP	enhanced system improvement program
EW	electronic warfare
FBCB2	Force XXI Battle Command Brigade and Below
FCTN	function
FEC	forward error correction
FH	frequency hopping
FHMUX	frequency hopping multiplexer
FLOT	forward line of own troops
FM	field manual; frequency modulation
FMI	field manual interim
FSK	frequency shift key
ft	feet
G-2	assistant chief of staff, intelligence
G-3	assistant chief of staff, operations
G-6	assistant chief of staff, command, control, communications, and computer operations
GCC	geographic combatant commander
GHz	gigahertz
GPS	global positioning system
HDR	high data rate
HF	high frequency
HRCRD	handheld remote control radio device
HQ	headquarters
Hz	hertz
I/O	input/output
ICEPAC	Ionospheric Communications Enhanced Profile and Circuit
ICOM	integrated communications security

IED	improvised explosive device
IHFR	improved high frequency radio
IMMP	Information Management Master Plan
IMP	Information Management Plan
INC	internet controller card
INMARSAT	international maritime satellite
IONCAP	Ionospheric Communications Analysis and Prediction
IP	Internet Protocol
ISP	Information Systems Plan
ISSO	Information Systems Security Officer
ITSB	integrated theater signal battalion
IVRCU	intravehicular remote control unit
J-6	communications system directorate of a joint staff
JACS	joint automated communications-electronics operating instructions system
JCF	joint contingency force
JCS	Joint Chiefs of Staff
JINTACCS	Joint Interoperability of Tactical Command and Control System
JLENS	Joint Land Attack Cruise Missile Defense Elevated Netted Sensor System
JNN	Joint Network Node
JP	joint publication
JRFL	joint restricted frequency list
JSIR	Joint Spectrum Interference Resolution
JTAGS	joint tactical ground station
JTF	joint task force
JTIDS	Joint Tactical Information Distribution System
JTRS	Joint Tactical Radio System
kbps	kilobits per second
KEK	key encryption key
kHz	kilohertz
km	kilometer
KMP	key management plan
kW	kilowatt
LCMS	local communications security management software
LCU	lightweight computer unit
LDV	last ditch voice
LMR	land mobile radio
LOS	line of sight
LPI/D	low probability of interception/detection
LQA	link quality analysis

LRU	line replaceable unit
LTS	logical time slot
LVT	low volume terminal
LW	Land Warrior
MBITR	multiband inter/intra team radio
MCS	master control station
MEADS	medium extended air defense system
MELP	mixed excitation linear prediction
METT-TC	mission, enemy, terrain and weather, troops and support available, time available, civil considerations
MF	medium frequency
MHz	megahertz
MIDS	Multifunctional Information Distribution System
MIL-STD	military standard
MLRS	Multiple Launch Rocket System
MNL	master net list
MSE	mobile subscriber equipment
MSG	multisource group
MSRT	mobile subscriber radiotelephone terminal
MUF	maximum usable frequency
NATO	North Atlantic Treaty Organization
NCO	noncommisioned officer
NCS	net control station
net	network
NET ID	network identifier
NMS	Network Management System
NSA	National Security Agency
NTDR	near term digital radio
NTIA	National Telecommunications and Information Administration
NVIS	near-vertical incident sky wave
O&I	operations and intelligence
OCONUS	outside the continental United States
OPLAN	operation plan
OPORD	operation order
OTAR	over-the-air rekeying
P25	Project 25 standards (APCO)
PA	power amplifier
PC	personal computer
PLGR	precision lightweight global positioning system receiver
PMCS	preventive maintenance checks and services
PSK	phase shift keying

PSYOP	Psychological Operations
PT	plain text
QEAM	quick erect antenna mast
RBECS	Revised Battlefield Electronic Communications-Electronics Operational Instruction/Signal Operating Instructions System
RCU	remote control unit
RDG	random data generator
RDS	revised data transfer device software
RF	radio frequency
RSA	radio set adapter
RT	receiver/transmitter
RTO	radio-telephone operator
S-2	intelligence staff officer
S-3	operations staff officer
S-6	signal staff officer
SA	situational awareness
SAMS-1	Standard Army Maintenance System-Level 1
SATCOM	satellite communications
SC	single-channel
SCAMP	single-channel anti-jam man portable
SC TACSAT	single-channel tactical satellite
SECOMP	secure en route communications package
SFAF	standard frequency action format
SHORAD	short-range air defense
SIGINT	signals intelligence
SINAD	signal, noise and distortion
SINCGARS	Single-Channel Ground and Airborne Radio System
SIP	system improvement program
SKL	simple key loader
S/N	signal to noise
SOF	Special Operations Forces
SOI	signal operating instructions
SOP	standing operating procedure
SPEED	system planning, engineering, and evaluation device
SQ	squelch
SRCU	securable remote control unit
SSB	single side band
STANAG	standardization agreement (NATO)
STU	secure telephone unit
TAC CP	tactical command post
TACFIRE	Tactical Fire Direction System

TACSAT	tactical satellite
TADIL-J	tactical digital information link-joint
TAOM	Tactical Air Operations Module
TB	technical bulletin
TDMA	time division multiple access
TEK	traffic encryption key
THAAD	Theater High Altitude Air Defense
TM	technical manual
TOC	tactical operations center
TRADOC	United States Army Training and Doctrine Command
TRANSEC	transmission security
TSEC	telecommunications security
TSK	transmission security key
TTP	tactics, techniques, and procedures
TTSB	tactical theater signal brigade
UHF	ultra high frequency
US	United States
US&P	United States and Possessions
USAF	United States Air Force
USAR	United States Army Reserve
USMC	United States Marine Corps
USN	United States Navy
VAA	vehicular amplifier adapter
VAC	volts alternating current
VDC	volts direct current
VHF	very high frequency
VIS	vehicular intercommunications system
VOACAP	Voice of America Coverage Analysis Program
VOCODER	voice encoder
VSWR	voltage standing wave radio
WIN-T	Warfighter Information Network-Tactical

SECTION II – TERMS

*acknowledge

A directive from the originator of a communication requiring the address(s) to advise the originator that his communication has been received and understood. This term is normally included in the electronic transmission of orders to ensure the receiving station or person confirms the receipt of the order.

*all after

A procedure word meaning, "The portion of the message to which I have referenced is all that follows (insert text)." See also **procedure word**.

*all before

A procedure word meaning, "The portion of the message to which I have reference is all that precedes (insert text)." See also **procedure word**.

area of operations

(joint) An operational area defined by the joint force commander for land and maritime forces. Areas of operations do not typically encompass the entire operational area of the joint force commander, but should be large enough for component commanders to accomplish their missions and protect their forces. (JP 3-0)

*authenticate

A procedure word meaning, "The station called is to reply to the challenge which follows (insert text)".

*authentication

A security measure designed to protect a communications system against acceptance of a fraudulent transmission or simulation by establishing the validity of a transmission, message, or originator.

*authentication is

A procedure word meaning, " The transmission authentication of this message is " (insert text)." See also **procedure word**.

azimuth

(joint) Quantities may be expressed in positive quantities increasing in a clockwise direction, or in X, Y coordinates where south and west are negative. They may be referenced to true north or magnetic north depending on the particular weapon system used.

bandwidth

(joint) The difference between the limiting frequencies of a continuous frequency expressed in hertz (cycles per second). The term bandwidth is also loosely used to refer to the rate at which data can be transmitted over a given communications circuit. In the latter usage, bandwidth is usually expressed in either kilobits per second or megabits per second. (JP 1-02)

beam width

(joint) The angle between the directions on either side of the axis, at which the intensity of the radio frequency field drops to one-half the value it has on the axis. (JP 1-02)

*break

A procedure word meaning, "I here by indicate the separation of the text from another portion of the message." See also **procedure word**.

call sign

(joint) Any combination of characters or pronounceable words, which identifies a communication facility, a command, an authority, an activity, or a unit; used primarily for establishing and maintaining communications. (JP 1-02)

*clear

To eliminate transmissions on a tactical radio net in order to allow a higher-precedence transmission to occur.

command and control

(Army) The exercise of authority and direction by a properly designated commander over assigned and attached forces in the accomplishment of a mission. Commanders perform command and control functions through a command and control system. (FM 6-0)

command post

(Army) A unit's or subunit's headquarters where the commander and the staff perform their activities. (FM 6-0)

common operating environment

(joint) Automation services that support the development of the common reusable software modules that enable interoperability across multiple combat support applications. This includes segmentation of common software modules from existing applications, integration of commercial products, development of a common architecture, and development of common tools for application developers. (JP 4-01)

communications net

(joint) An organization of stations capable of direct communications on a common channel or frequency. (JP-1.02)

communications intelligence

(joint) Technical information and intelligence derived from foreign communications by other than the intended recipients. (JP 2-0)

communications security

(joint) The protection resulting from all measures designed to deny unauthorized persons information of value that might be derived from the possession and study of telecommunications, or to mislead unauthorized persons in their interpretation of the results of such possession and study. (JP 6-0)

*correct

A procedure word meaning, "You are correct, or what you have transmitted is correct". See also **procedure word**.

*correction

A procedure word meaning, 1. "An error has been made in this transmission. Transmission will continue with the last word correctly transmitted." 2. "An error has been made in thes transmission (or message indicated). The correct version is (insert text)." 3. "That which follows is a corrected version in answer to your request for verification." See also **procedure word**.

*disregard this transmission-out

A procedure word meaning, "This transmission is in error. Disregard it." (This procedure word shall not be used to cancel any message that has been completely transmitted and for which a receipt or acknowledgement has been received.) See also **procedure word**.

*do not answer

A procedure word meaning, "Stations called are not to answer this call, receipt for this message, or otherwise transmit in connection with this transmission." When this procedure word is employed, the transmission shall be ended with the procedure word "Out." See also **procedure word**.

electromagnetic interference

(joint) Any electromagnetic disturbance that interrupts, obstructs, or otherwise degrades or limits the effective performance of electronics and electrical equipment. It can be induced intentionally, as in some forms of electronic warfare, or unintentionally, as a result of spurious emissions and responses, intermodulation products, and the like. (JP 1-02)

electromagnetic pulse

(joint) The electromagnetic radiation from a strong electronic pulse, most commonly caused by a nuclear explosion, that may couple with electrical or electronic systems to produce damaging current and voltage surges. (JP 3-13.1)

electromagnetic spectrum

(joint) The range of frequencies of electromagnetic radiation from zero to infinity. It is divided into 26 alphabetically designated bands. (JP 1-02)

electronic protection

(joint) Division of electronic warfare involving actions taken to protect personnel, facilities, and equipment from any effects of friendly or enemy use of the electromagnetic spectrum that degrade, neutralize or destroy friendly combat capability. (JP 3-13.1)

electronic warfare

(joint) Military action involving the use of electromagnetic and directed energy to control the electromagnetic spectrum or to attack the enemy. Electronic warfare consists of three divisions: electronic attack, electronic protection, and electronic warfare support. (JP 3-13.1)

electronic warfare support

(joint) Division of electronic warfare involving actions tasked by, or under direct control of, an operational commander to search for, intercept, identify, and locate or localize sources of intentional and unintentional radiated electromagnetic energy for the purpose of immediate threat recognition, targeting, planning, and conduct of future operations. (JP 3-13.1)

emission control

(joint) The selective and controlled use of electromagnetic, acoustic, or other emitters to optimize command and control capabilities while minimizing, for operations security: a. detection by enemy sensors; b. mutual interference among friendly systems; and/or c. enemy interference with the ability to execute a military deception plan. (JP 1-02)

*exempt

A procedure word meaning, "The addresses immediately following are exempt from the collective call." See also **procedure word**.

*figures

A procedure word meaning, "Numerals or numbers follow." See also **procedure word**.

*flash

A procedure word meaning, "Precedence, FLASH." Reserved for initial enemy contact reports on special emergency operational combat traffic originated by specifically designated high commanders or units directly affected. This traffic is to be SHORT reports of emergency situations of vital proportions. Handling is as fast as humanely possible with an objective time of 10 minute or less. See also **procedure word**.

forward line of own troops

(joint) A line that indicates the most forward positions of friendly forces in any kind of military operation at a specific time. The forward line of own troops normally identifies the forward location of covering and screening forces. The FLOT may be at, beyond, or short of the forward edge of the battle area. An enemy FLOT indicates the forward-most position of hostile forces. (JP 1-02)

*from

A procedure word meaning, "The originator of this message is indicated by the address designator immediately following." See also **procedure word**.

*groups

A procedure word meaning, "This message contains the number of groups indicated." See also **procedure word**.

*guard

A procedure word meaning, "A...MHz" radio frequency that is normally used for emergency transmissions and is continuously monitored. UHF band: 243. MHz; VHF band: 121.5 MHz. See also **procedure word**.

*I authenticate

A procedure word meaning, "The group that follows is the reply to your challenge to authenticate." See also **procedure word**.

*immediate

A procedure word meaning, "Precedence immediate." The precedence reserved for messages relating to situations which gravely affect the security of national/multinational forces or populace, and which require immediate delivery. See also **procedure word**.

*info

A procedure word meaning, "The addressees immediately following are addressed for information." See also **procedure word**.

information superiority

(joint) The operational advantage derived from the ability to collect, process, and disseminate an uninterrupted flow of information while exploiting or denying an adversary's ability to do the same. (JP 3-13)

*I read back

A procedure word meaning, "The following is my response to your instructions to read back." See also **procedure word**.

*I say again

A procedure word meaning, "I am repeating transmission or portion indicated." See also **procedure word**.

*I spell

A procedure word meaning, "I shall spell the next word phonetically." See also **procedure word**.

*I verify

A procedure word meaning, "That which follows has been verified at your request and is repeated." (to be used as a reply to "verify"). See also **procedure word**.

jamming

(Army) The deliberate radiation or reflection of electromagnetic energy to prevent or degrade the receipt of information by a receiver. It includes communications and noncommunications jamming. (FM 2-0)

line of sight

(Army) The unobstructed path from a Soldier/Marine, weapon, weapon sight, electronic-sending and -receiving antennas, or piece of reconnaissance equipment to another point. (FM 34-130)

man portable

(joint) Capable of being carried by one man. Specifically, the term may be used to qualify 1. Items designed to be carried as an integral part of individual, crew-served, or team equipment of the dismounted Soldier in conjunction with assigned duties. Upper weight limit: approximately 14 kilograms (31 pounds). 2. In land warfare, equipment which can be carried by one man over long distance without serious degradation of the performance of normal duties. (JP 1-02)

*message

A procedure word meaning, "A message which requires recording is about to follow." See also **procedure word**.

*more to follow

A procedure word meaning "Transmitting station has additional traffic for the receiving station." See also **procedure word**.

multichannel

(joint) Pertaining to communications, usually full duplex, on more than one channel simultaneously. Multichannel transmission may be accomplished by either time-, frequency-, code-, and phase-division multiplexing or space diversity. (JP 1-02)

near real time

(joint) Pertaining to the timeliness of data or information which has been delayed by the time required for electronic communication and automatic data processing. This implies that there are no significant delays. (JP 1-02)

net (communications)

(joint) An organization of stations capable of direct communications on a common channel or frequency. (JP 1-02)

net call sign

(joint) A call sign which represents all stations within a net. (JP 1-02)

*net control station

A communications station designated to control traffic and enforce circuit discipline within a given net.

operational environment

(joint) A composite of the conditions, circumstances, and influences which affect the employment of military forces and bear on the decisions of the unit commander. (JP 3-0)

*out

A procedure word meaning, "This is the end of my transmission to you and no answer is required or expected." (Since "over" and "out" have opposite meanings, they are never used together.) See also **procedure word**.

*over

A procedure word meaning, "This is the end of my transmission to you and a response is necessary. Go ahead; transmit." See also **procedure word**.

phonetic alphabet

(joint) A list of standard words used to identify letters in a message transmitted by radio or telephone. (JP 1-02)

*priority

A procedure word meaning, "Precedence priority." Reserved for important messages that must have precedence over routine traffic. This is the highest precedence that normally may be assigned to a message of administrative nature. See also **procedure word**.

*procedure word

A word or phrase limited to radio telephone procedure used to facilitate communication by conveying information in a condensed standard form. Also called **proword**.

*radio listening silence

The situation where radios are on and continuously monitored with strict criteria when a station on the radio network is allowed to break silence. For example, "maintain radio listening silence until physical contact with the enemy is made."

*read back

A procedure word meaning, "Repeat this entire transmission back to me exactly as received." See also **procedure word**.

*relay to

A procedure word meaning, "Transmit this message to all addressees (or addressees immediately following this proword)." The address component is mandatory when this proword is used. See also **procedure word**.

*roger

A procedure word meaning "I have received your last transmission satisfactorily." See also **procedure word**.

*routine

A procedure word meaning, "Precedence routine." Reserved for all types of messages that are not of sufficient urgency to justify a higher precedence, but must be delivered to the addressee without delay. See also **procedure word**.

*say again

A procedure word meaning, "Repeat all of your last transmission." (Followed by identification data, means "Repeat _____ (portion indicated).") See also **procedure word**.

SECRET Internet Protocol Router Network

(joint) The worldwide SECRET-level packet switch network that uses high-speed Internet protocol routers and high-capacity Defense Information Systems Network circuitry. (JP 6-0)

signal

(joint) 1. As applied to electronics, any transmitted electrical impulse. 2. Operationally, a type of message, the text of which consists of one or more letters, words, characters, signal flags, visual displays, or special sounds with prearranged meaning, and which is conveyed or transmitted by visual, acoustic, or electrical means. (JP 1-02)

*signal operating instructions

A series of orders issued for technical control and coordination of the signal communication activities of a command.

*signal security

A generic term that includes both communications security and electronics security. Measures intended to deny or counter hostile exploitation of electronic emissions. Signal security includes communications security and electronic security.

signal to noise ratio

(joint) The ratio of the amplitude of the desired signal to the amplitude of noise signals at a given point in time. (JP 1-02)

*silence

A procedure word meaning, "Cease transmission immediately." Silence will be maintained until lifted. (Transmissions imposing silence must be authenticated.) See also **procedure word**.

*silence lifted

A procedure word meaning, "Silence is lifted." (When an authentication system is in force, the transmission lifting silence is to be authenticated.) See also **procedure word**.

SIPRNET

See SECRET Internet Protocol Router Network.

*speak slower

A procedure word meaning, "Your transmission is at too fast a speed. Reduce speed of transmission." See also **procedure word**.

TABOO frequencies

(joint) Any friendly frequency of such importance that it must never be deliberately jammed or interfered with by friendly forces. Normally, these frequencies include international distress, CEASE BUZZER, safety, and controller frequencies. These frequencies are generally long standing. However, they may be time-oriented in that, as the conduct or exercise situation changes, the restrictions may be removed. (JP 1-02)

tactical call sign

(joint) A call sign which identifies a tactical command or tactical communication facility. (JP 1-02)

*this is

A procedure word meaning, "This transmission is from the station whose designator immediately follows." See also **procedure word**.

*time

A procedure word meaning, "That which immediately follows is the time or date/time group of the message." See also **procedure word**.

*to

A procedure word meaning, "The addressee(s) immediately following is (are) addressed for action." See also **procedure word**.

transponder

(joint) A receiver-transmitter which will generate a reply signal upon proper interrogation. (JP 1-02)

Universal Time

(joint) A measure of time that conforms, within a close approximation, to the mean diurnal rotation of the Earth and serves as the basis of civil timekeeping. Universal Time (UT1) is determined from observations of the stars, radio sources, and also from ranging observations of the moon and artificial Earth satellites. The scale determined directly from such observations is designated Universal Time Observed (UTO); it is slightly dependent on the place of observation. When UTO is corrected for the shift in longitude of the observing station caused by polar motion, the time scale UT1 is obtained. When an accuracy better than one second is not required, Universal Time can be used to mean Coordinated Universal Time. (JP 1-02)

*unknown station

A procedure word meaning, "The identity of the station with whom I am attempting to establish communications is unknown." See also **procedure word**.

urban operations

(Army) Offense, defense, stability, and support operations conducted in a topographical complex and adjacent natural terrain where manmade construction and high population density are the dominant features. (FM 3-0)

*verify

A procedure word meaning, "Verify entire message (or portion indicated) with the originator and send correct version." (To be used only at the discretion of the addressee to which question message was directed.) See also **procedure word**.

*wait

A procedure word meaning, "I must pause for a few seconds." See also **procedure word**.

*wait out

A procedure word meaning, "I must wait for longer than a few seconds." See also **procedure word**.

way point

(joint) A designated point or series of points loaded and stored in a global positioning system or other electronic navigational aid system to facilitate movement. (JP 1-02)

*wilco

A procedure word meaning, "I have received your signal, understand it, and will comply." (To be used only by addressee. Since the meaning of ROGER is included in that of WILCO, the two procedure words are never used together.) See also **procedure word**.

*word after

A procedure word meaning, "The word of the message to which I have reference is that which follows (insert text)." See also **procedure word**.

*word before

A procedure word meaning, "The word of the message to which I have reference is that which precedes (insert text)." See also **procedure word**.

*words twice

A procedure word meaning, "Communication is difficult. Transmit (ring) each phrase (or each code group) twice." This procedure word may be used as an order, request, or as information. See also **procedure word**.

*wrong

A procedure word meaning, "Your last transmission was incorrect, the correct version is (insert text)." See also **procedure word**.

Zulu Time

See Universal Time.

References

SOURCES USED

These are the sources quoted or paraphrased in this publication. Allied publications can be found online at http://www.jcs.dtic.mil/j6/cceb/acps/. Most Army doctrinal publications are available online at https://akocomm.us.army.mil/usapa/. (Access requires an Army Knowledge Online account.) Most joint publications can be found online at http://www.dtic.mil/doctrine/jpcsystemsseriespubs.htm. Publications from the *Army Communicator* can be found online at http://www.gordon.army.mil/ac/default.asp. Other useful Web sites are http://www.gordon.army.mil/sigbde15/25U/FAQ.htm and https://lwneusignal.army.mil/login.html.

ALLIED COMMUNICATIONS PUBLICATIONS

ACP 121 (H). *Communications Instructions General*. April 2007.

ACP 125 US SUPP-l. *Communications Instructions Radiotelephone Procedures for Use by United States Ground Forces*. October 1985.

JOINT PUBLICATIONS

CJCSM 3320.02B. *Joint Spectrum Interference Resolution (JSIR) Procedures*. 31 December 2008.

CJCSI 6251.01B. *Ultrahigh Frequency (UHF) Satellite Communications Demand Assigned Multiple Access Requirements*. 20 November 2007.

JP 1-02. *Department of Defense Dictionary of Military Terms and Associated Terms*. 12 April 2001.

JP 3-13.1 *Electronic Warfare*. 25 January 2007.

JP 6-0. *Joint Communications Systems*. 20 March 2006.

MIL-STD-196E. *Joint Electronics Type Designator System*. 17 February 1998.

Manual of Regulations and Procedures for Federal Radio Frequency Management (NTIA Manual). US Department of Commerce, NTIA. January 2009. http://www.ntia.doc.gov/osmhome/redbook/redbook.html.

ARMY PUBLICATIONS

AR 5-12. *Army Management of the Electromagnetic Spectrum*. 01 October 1997.

AR 25-2. *Information Assurance*. 24 October 2007.

AR 70-38. *Research, Development, Test and Evaluation of Material for Extreme Climate Conditions*. 15 September 1979.

AR 380-5. *Department of the Army Information Security Program*. 29 September 2000.

AR 380-40. *Policy for Safeguarding and Controlling Communications Security (COMSEC) Material*. 30 June 2000.

AR 380-53. *Information System Security Monitoring*. 29 April 1998.

FM 1-02. *Operational Terms and Graphics*. 21 September 2004.

FM 1-02.1 (FM 3-54.10). *Multi-Service Brevity Codes*. 30 October 2007.

FM 2-0 (FM 34-1). *Intelligence*. 17 May 2004.

FM 3-0. *Operations*. 27 February 2008.

FM 3-04.111. *Aviation Brigades*. 07 December 2007.

FM 3-09.21 (FM 6-20-1). *Tactics, Techniques and Procedures for the Field Artillery Battalion*. 22 March 2001.

FM 3-25.26. *Map Reading and Land Navigation*. 18 January 2005.

FM 6-02.72. *Multiservice Communications Procedures for Tactical Radios in a Joint Environment*. 14 June 2002.

FM 6-02.74. *Multi-Service Tactics, Techniques, and Procedures for High Frequency-Automatic Link Establishment (HF ALE) Radios*. 20 November 2007.

FM 6-02.771. *Multiservice Tactics, Techniques, and Procedures for Have Quick Radios*. 07 May 2004.

FM 6-02.90. *Multi-service Tactics, Techniques, and Procedures for Ultra High Frequency Tactical Demand Assigned Multiple Access Operations*. 31 August 2004.

FM 6-50. Tactics, Techniques, and Procedures for the Field Artillery Cannon Battery. 23 December 1996.

FM 7-0. *Training for Full Spectrum Operations*. 12 December 2007.

TB 11-5820-890-12. *Operator and Unit Maintenance for AN/CYZ-10 Automated Net Control Device (ANCD) (NSN 5810-01-343-1194) (EIC: QSU) with the Single Channel Ground and Airborne Radio Systems (SINCGARS)(AR)*. 01 April 1993.

TB 11-5820-1171-10. Software User's Guide for Near Term Digital Radio (NTDR) Network Management Terminal (NMT) (NSN: N/A) (EIC: N/A). 01 May 2005.

TB 11-5820-1172-10. *Operator and Maintenance Manual for Defense Advanced GPS Receiver (DAGR) Satellite Signals Navigation Set AN/PSN-13 AN/PSN-13 (NSN 5825-01-516-8038) AN/PSN-13A (NSN 5825-01-526-4783)*. 01 March 2005.

TB 11-5821-333-10-2. *SINCGARS Airborne ICOM Radio Operator's Pocket Guide, SINCGARS Airborne ICOM Radios used with Automated Net Control Device, AN/CYZ-10*. 01 July 1995.

TB 11-5825-291-10-2. *Soldiers Guide for Precision Lightweight GPS Receiver (PLGR) AN/PSN-11 (NSN 5825-01-374-6643) (EIC: N/A) and AN/PSN-11(v) (5825-01-395-3513) (EIC: N/A)*. 01 December 1996.

TB 11-5825-298-10-1. *Operator's Manual for Net Control Station AN/TSQ-158A (NSN 5895-01-495-5977) (EIC: N/A) Part of Enhanced Position Location Reporting System (EPLRS)*. 01 May 2005.

TB 11-7010-293-10-2. *Operator's Manual Automated Communications Engineering Software (ACES) Version 1.9 for AN/GYK-33D (NSN: 7010-01-541-5396) (EIC; N/A)*. 01 June 2009.

TB 380-41. *Security: Procedures for Safeguarding, Accounting, and Supply control of COMSEC Material*. 15 March 2006.

TC 2-33.4 (FM 34-3). *Intelligence Analysis*. 01 July 2999.

TC 9-64. *Communications-Electronics Fundamentals: Wave Propagation, Transmission Lines, and Antennas*.15 July 2004.

TM 11-5820-890-10-5. *SINCGARS Icom and Non-Icom Ground Radio Net Control Station (NCS) Pocket Guide Radio Set Manpack Radio (AN/PRC-119/119A) Vehicular Radios (AN/VRC-87/87A-C Thru AN/VRC-92/92A)*. 01 April 1993.

TM 11-5820-890-10-6. *SINCGARS Icom Ground Radios Used With Automated Net Control Device (ANCD) AN/CYZ-10; Precision Lightweight GPS Receiver (PLGR) AN/PSN-11; Handheld Remote Control Radio Device (HRCRD)C-12493/U; Simple Key Loader (SKL) AN/PYQ-10; Operator's Pocket Guide Radio Manpack Radios (AN/PRC-119A/D/F) (NSN: N/A) (EIC: N/A) Vehicular Radios (AN/VRC-87A/D/F THRU AN/VRC-92A/D/F) (NSN: N/A) (EIC: N/A)*. 01 July 2007.

TM 11-5820-890-10-7. *SINCGARS Icom Ground Radios Used With Automated Net Control Device (ANCD) AN/CYZ-10, Precision Lightweight GPS Receiver (PLGR) AN/PSN-11 Handheld Remote Control Radio Device (HRCRD) C-12493/U; Simple Key Loader (SKL) AN/PYQ-10 Net Control Station (NCS) Pocket Guide Manpack Radios AN/PRC-119A/D/F (NSN: N/A) (EIC: N/A) Vehicular Radios AN/VRC-87A/D/F Thru AN/VRC-92A/D/F (NSN: N/A)(EIC: N/A)*. 01 August 2007.

TM 11-5820-890-10-8. *Operator's Manual for SINCGARS Ground Combat Net Radio, ICOM Manpack Radio AN/PRC-119A (NSN 5820-01-267-9482) (EIC: L2Q), Short Range Vehicular Radio AN/VRC-87A (5820-01-267-9480) (EIC: L22), Short Range Vehicular Radio with Single Radio Mount AN/VRC-87C (5820-01-304-2045) (EIC: GDC), Short Range Vehicular Radio with*

Dismount AN/VRC-88A (5820-01-267-9481) (EIC: L23), Short Range/Long Range Vehicular Radio AN/VRC-89A (5820-01-267-9479) (EIC: L24), Long Range Vehicular Radio AN/VRC-90A (5820-01-268-5105) (EIC: L25), Short Range/Long Range Vehicular Radio with Dismount AN/VRC-91A (5820-01-267-9478) (EIC: L26), Short Range/Long Range Vehicular Radio AN/VRC-92A (5820-01-267-9477) (EIC: L27) Used With Automated Net Control Device (ANCD) (AN/CYZ-10) Precision Lightweight GPS Receiver (PLGR) (AN/PSN-11) Secure Telephone Unit (STU) Frequency Hopping Mutiplexer (FHMUX). 01 December 1998.

TM 11-5820-890-23P. *Unit and Direct Support Maintenance Repair Parts and Special Tools List for FHMUX TD-1456/VRC (NSN 5820-01-365-2721) (EIC: N/A) Mount MT-6845/VRC (5975-01-430-3109) (EIC: N/A).* 01 October 1998.

TM 11-5820-919-12. *Operator's and Organizational Maintenance Manual for Radio Set, AN/PRC-104A (NSN 5820-01-141-7953).* 15 January 1986.

TM 11-5820-923-12. *Operator's and Organizational Maintenance Manual for Radio Set, AN/GRC-213 (NSN 5820-01-128-3935).* 14 February 1986.

TM 11-5820-924-13. *Operator's, Organizational and Direct Support Maintenance Manual for Radio Set, AN/GRC-193A (NSN 5820-01-133-4195).* 14 February 1986.

TM 11-5820-1025-10. *Operator's Manual for Radio Set, AN/PRC-126 (NSN 5820-01-215-6181).* 01 February 1988.

TM 11-5820-1037-13&P. *Operator's, Unit, and Intermediate Maintenance Manual (Repair Parts and Special Tools List) for Radio Set AN/PRC-112 (NSN 5820-01-279-5450) (EIC: JBG) Program Loader KY-913/PRC-112 (NSN 7025-01-279-5308) (EIC: N/A).* 15 July 2005.

TM 11-5820-1049-12. *Operator's and Aviation Unit Maintenance Manual for Radio Set AN/PRC-90-2 (NSN 5820-01-238-6603).* 15 August 1990.

TM 11-5820-1130-12&P. *Operator's and Unit Maintenance Manual (Including Repair Parts and Special Tools List) for Radio Set AN/PSC-5 (NSN 5820-01-366-4120) (EIC: N/A).* 01 June 2000.

TM 11-5820-1141-12&P. *Operator and Unit Maintenance Manual (Including Repair Parts and Special Tools List) for Radio Set AN/VRC-100(V)1 (NSN: 5820-01-413-4235) (EIC: N/A).* 01 December 2004.

TM 11-5820-1149-14&P. *Operator's Unit, Direct and General Support Maintenance Manual (Including Repair Parts and Special Tools List) for Radio Set AN/VRC-83(V)3 (NSN 5820-01-291-5415) (EIC: N/A).* 01 April 1996.

TM 11-5820-1157-10. *Operator's Manual for AN/PSC-11 Single Channel Anti-Jam Manportable (SCAMP) Terminal (NSN 5820-01-431-2060) (EIC: N/A).* 01 May 2003.

TM 11-5820-1171-12&P. *Operator's and Unit Maintenance Including Repair Parts and Special Tools List for Radio Set AN/VRC-108 (Near Term Digital Radio (NTDR)) (NSN 5820-01-519-2729) (EIC:N/A).* 01 May 2005.

TM 11-5820-1172-13. *Operator and Maintenance Manual Defense Advanced GPS Receiver (DAGR) Satellite Signals Navigation Set AN/PSN-13 (NSN 5825-01-516-8038) AN/PSN-13A (NSN 5825-01-526-4783).* 01 March 2005.

TM 11-5821-318-12. *Operator's and Aviation Unit Maintenance Manual for VHF AM/FM Radio Set AN/ARC-186(V) (NSN 5821-01-086-6243) (EIC: N/A).* 01 September 2005.

TM 11-5821-333-12. *Operator's and Aviation Unit Maintenance Manual for SINCGARS Airborne Combat Net Radio, Icom and Non-Icom; Non-Icom Airborne Radio AN/ARC-201(V) (NSN: N/A) (EIC:N/A) Icom Airborne Radio AN/ARC-201A(V) (NSN: N/A) (EIC: N/A).* 01 September 1992.

TM 11-5821-357-12&P. *Operator's and Aviation Unit Maintenance Manual (Including Repair Parts and Sepcial Tools List) for Radio Set AN/ARC-220(V)1 (NSN 5821-01-413-4233) (EIC: GC6) and AN/ARC-220(V)2 (5821-01-413-4232) (EIC: GC7).* 01 June 2001.

TM 11-5825-283-10. *Operator's Manual for Manpack Radio Set (MP-RS) Radio Sets AN/ASQ-177C(V)4 (NSN 5820-01-462-8407) (EIC: N/A); AN/PSQ-6C (5820-01-462- 8410) (EC: N/A); AN/VSQ-2C(V)1 (5820-01-462-8411) (EIC: N/A): AN/VSQ-2C(V)2 (5820-01-462-8404) (EIC: N/A); AN/VSQ-2C(V)4 (5820-01-462-8408) (EIC: N/A); Grid Reference Radio Set AN/GRC-229C*

(5895-01-462-8405) (EIC: N/A); Downsized Enhanced Command Response Unit RT-1718/TSQ-158A (5820-01-381-6339) (EIC: N/A). 15 August 2000.

TM 11-5825-291-13. *Operations and Maintenance Manual for Satellite Signals Navigation Sets AN/PSN-11 (NSN 5825-01-374-6643) and AN/PSN-11(V)1 (5825-01-395-3513).* 01 April 2001.

TM 11-5825-298-13&P. *Operator and Field Maintenance Manual (Including Repair Parts and Special Tools List) for Net Control Station (NCS) AN/TSQ-158A (NSN 5895-01-495-5977) (EIC: N/A) Part of Enhanced Position Location Reporting System (EPLRS).* 01 October 2006.

TM 11-5830-263-10. *Operator's Manual for Vehicular Intercommunication Set AN/VIC-3(V), Including: Control Indicator CD-82/VRC (NSN 5895-01-382-3221) (EIC: N/A) Control Intercommunication Set C-12357/VRC (5830-01-382-3218) (EIC: N/A) Control Intercommunication SET C-12358/VRC (5830-01-382-3209) (EIC: N/A) Interface Unit, Communication Equipment C-12359/VRC (5895-01-382-3220) (EIC: N/A) Loudspeaker LS-688/VRC (5965-01-382-3222) (EIC: N/A).* 01 May 1997.

TM 11-5841-286-13. *Operator's, Organizational, and Direct Support Maintenance Manual: Radio Sets, AN/ARC-164(V)12 (NSN 5821-01-071-5624) AND AN/ARC-164(V)16 (5841-01-122-7094).* 30 July 1980.

TM 11-5985-357-13. *Operator's, Organizational, and Direct Support Maintenance Manual for Antenna Group, OE-254/GRC (NSN 5985-01-063-1574).* 01 February 1991.

MARINE CORP PUBLICATIONS

Marine Corps Reference Publication 622D. *Field Antenna Handbook.* 01 June 1999.

NONMILITARY PUBLICATIONS

Farmer, Edward. Long-range Communications at High Frequencies. *Army Communicator.* Winter, 2002.

Fiedler, David, M. AN/PRC-117F Special Operations Forces Radio has Applications For Digital Divisions and Beyond. *Army Communicator.* Summer, 2000.

Fiedler, David, M. MBITR Communications = Power in Your Pocket. *Army Communicator.* Summer, 2005.

Fiedler, David, M. Planning for the use of high-frequency radios in the brigade combat teams and other transformation Army organizations. *Army Communicator.* Fall, 2002.

Fiedler, David, M. Tactical Ground-Wave Communications for Force XXI Tactical Internet and Beyond. *Army Communicator.* Summer, 1996.

Fiedler, David, M. & Farmer, Edward. AN/PRC-150 HF Radio in Urban Combat: A Better Way To Command and Control the Urban Fight. *Army Communicator.* Spring, 2004.

Flynn, Mike. Fulfilling the Promise of HF. *Army Communicator.* Winter, 2007.

DOCUMENTS NEEDED

These documents must be available to the intended user of this publication.

DA Form 2028. *Recommended Changes to Publications and Blank Forms.*

Index

JOYCE E. MORROW
Administrative Assistant to the
Secretary of the Army
0919603

DISTRIBUTION:

Active Army, Army National Guard, and U.S. Army Reserve: Not to be distributed; electronic media only.

www.ingramcontent.com/pod-product-compliance
Lightning Source LLC
Chambersburg PA
CBHW051207200326
41519CB00025B/7038